Physarum Machines
Computers from Slime Mould

WORLD SCIENTIFIC SERIES ON NONLINEAR SCIENCE

Editor: Leon O. Chua
University of California, Berkeley

*To view the complete list of the published volumes in the series, please visit:
http://www.worldscibooks.com/series/wssnsa_series.shtml

WORLD SCIENTIFIC SERIES ON
NONLINEAR SCIENCE

Series A   Vol. 74

Series Editor: Leon O. Chua

Physarum Machines
Computers from Slime Mould

Andrew Adamatzky
University of the West of England, UK

World Scientific

NEW JERSEY · LONDON · SINGAPORE · BEIJING · SHANGHAI · HONG KONG · TAIPEI · CHENNAI

Published by

World Scientific Publishing Co. Pte. Ltd.

5 Toh Tuck Link, Singapore 596224

USA office: 27 Warren Street, Suite 401-402, Hackensack, NJ 07601

UK office: 57 Shelton Street, Covent Garden, London WC2H 9HE

British Library Cataloguing-in-Publication Data
A catalogue record for this book is available from the British Library.

World Scientific Series on Nonlinear Science, Series A — Vol. 74
PHYSARUM MACHINES
Computers from Slime Mould

ISBN-13 978-981-4327-58-9
ISBN-10 981-4327-58-1

Printed in Singapore.

Preface

Unconventional computing is chock full of theoretical stuff. There are a just a handful of experimental laboratory prototypes. They are outstanding but difficult for non-experts to play with. We show how to make a universal biological computer at home nearly for free. All you need is a slime mould, oat flakes and a camera. The rest is up to your creativity.

Plasmodium of the acellular slime mould *Polycephalum polycephalum* is our computing substrate. The plasmodium is a single monstrously large cell. The plasmodium behaves as a nonlinear medium, an excitable soft matter, encapsulated in an elastic and growing membrane.

In Chap. 1 we establish links between reaction–diffusion chemical computers and plasmodium of *P. polycephalum*. We also provide an insight on history of Physarum computing.

Where to get plasmodium? How to feed it? What substrates to use? Answers are given in Chap. 1.5.

Shortest path and plane tessellation are classical problems of computer science. In Chaps. 2.9 and 3.3 we show how to solve a maze and approximate a Voronoi diagram with plasmodium.

We aim not to clutter the book with theory. Thus, we provide just one simple yet illustrative model of Physarum behavior. We simulate propagating plasmodium fronts with a two-variable Oregonator. In Chap. 4.3 we use the Oregonator to uncover mechanics of spanning tree construction by Physarum.

When the plasmodium is placed on a substrate populated with sources of nutrients, it spans the sources with its protoplasmic network. The plasmodium optimizes the network to efficiently transport protoplasm with nutrients. How exactly does the protoplasmic network unfold during the plasmodium's foraging behavior? What types of proximity graphs are approximated by the network? How does the plasmodium increase the reliability and through capacity of the network? Does it construct a minimal spanning tree first and then add additional protoplasmic tubes? We answer these questions in Chap. 5.5.

The plasmodium propagates as a traveling localization — a compact wave fragment of protoplasm — on a non-nutrient substrate. The plasmodium localization travels in its originally pre-determined direction for a substantial period of time even when no gradient of chemo-attractants is present. In Chap. 6.7.4 we utilize this property of Physarum localizations to design a two-input two-output Boolean logic gate. A first gate produces a disjunction of inputs on one output and a conjunction of inputs on another output. A second gate produces an unchanged input variable on one of the outputs. On the other output the gate brings out a conjunction of one input with the negation of another input. We cascade the logical gates into a one-bit half-adder and simulate its functionality.

A Kolmogorov–Uspensky machine and a Turing machine are birth cohorts. The Turing machine was crowned by theoreticians. The Kolmogorov–Uspensky machine became a prototype of modern computer architectures. In Chap. 7.7, we show how to imitate a Kolmogorov–Uspensky machine in plasmodium. A Physarum machine is an experimental implementation, modification and extension of a Kolmogorov–Uspensky machine in plasmodium of *P. polycephalum*. Data are represented by sources of nutrients and memory structure by protoplasmic tubes connecting the sources. By experimentally implementing the Kolmogorov–Uspensky machine in Physarum, we prove that plasmodium of *P. polycephalum* is a general-purpose computer.

Programming and reconfiguration of Physarum machines, and signal routing in Physarum circuits, are developed in Chaps. 8.4–10.10. We show how to program Physarum machines with attracting fields generated by sources of nutrients (Chap. 8.4). We demonstrate how to route signals in Physarum machines with domains of light (Chap. 9.4) and diffusive fields of repellents (Chap. 10.10).

Being encapsulated in an elastic membrane, the plasmodium can be capable of not only computing over spatially distributed datasets but also physically manipulating elements of the datasets. If a sensible, controllable and, ideally, programmable movement of the plasmodium was achieved, we would get experimental implementations of amorphous robotic devices.

Chapters 11.6–13.9 lay the experimental background for future plasmodium-based robotic devices. There we provide experimental evidence that plasmodium can sensibly manipulate tiny lightweight objects floating on a water surface (Chap. 11.6), act as an on-board 'motor' by propelling floaters with its oscillating pseudopodia (Chap. 12.3) and transport, mix and transform substances (Chap. 13.9).

Chapter 14.5 pushes the limits of our imagination a bit further. We allow the plasmodium to become a road-planning engineer. The structure of the protoplasmic networks, developed by the plasmodium, allows the plasmodium to optimize transfer of nutrients between remote parts of its body, to distributively sense its environment and to make a decentralized decision about further routes of migration. We consider the 10 most populated urban areas in the United Kingdom and study what would be an optimal layout of transport links between these urban areas from the 'plasmodium's point of view'.

A pocket manifest on Physarum machines is given in the epilogue.

Reader beware

> "...the girl I described, she's as real as the wind
> It's true I saw her today
> The other details are inventions
> Because I prefer her that way."

Bret McKenzie and Jemaine Clement (2009)[1]

Unconventional computing is an art of interpretation. Physarum does not compute. It obeys physical, chemical and biological laws. We translate its behavior to the language of computation. Reader beware. Our experiments are unbiased. Inferences might be slanted.

Andrew Adamatzky

---

[1]Bret McKenzie and Jemaine Clement. Rambling Through the Avenues of Time. Flight of the Conchords. October 2009.

Acknowledgments

It all started in 2005 when Soichiro Tsuda posted a sclerotium of *Physarum polycephalum* to me. "Try it, mate", he told me. I tried it and got hooked. Soichiro, thank you. Without you the book would never have been born.

Jeff Jones, your constant support, analysis and comments are priceless. We have done a lot together. You helped me a lot.

Tomohiro Shirakawa, Martin Grube, Ioannis Ieropoulos, Ben De Lacy Costello, Genaro Martinez, Andrew Schumann, Ramon Alonso-Sanz, Jonathan Mills, Larry Bull, Tetsuya Asai, Christof Teuscher and Toshiyuki Nakagaki, thank you guys, for your invisible contributions.

Ideas of amorphous robotic devices originate from my early collaboration with Chris Melhuish, Owen Holland and Hiroshi Yokoi. Guys, I am grateful!

My thanks go to Selim Akl for supporting my thoughts on Physarum machines and proximity graphs.

I am in debt to Leon O. Chua. Thank you, Leon, for your patronage, encouragement and inspiration.

Many thanks to Michael Jones for editing the manuscript.

Contents

Chapter 1

From reaction–diffusion to Physarum computing

Research in unconventional, or nature-inspired, computing aims to uncover novel principles of efficient information processing and computation in physical, chemical and biological systems, to develop novel non-standard algorithms and computing architectures, and also to implement conventional algorithms in non-silicon, or wet, substrates [Teuscher and Adamatzky (2005); Adamatzky and Teuscher (2006); Calude et al. (2006); Adamatzky et al. (2007)]. This emerging field of science and engineering is predominantly occupied by purely theoretical research, e.g. quantum computation, membrane computing and dynamical systems computing. Only a handful of experimental prototypes are reported so far, for example

- specialized and universal chemical reaction–diffusion processors [Adamatzky et al. (2005)],
- universal extended analog computers [Mills (2008)],
- maze-solving micro-fluidic circuits [Fuerstman et al. (2003)],
- gas-discharge analog path finders [Reyes et al. (2002)],
- maze-solving chemo-tactic droplets [Lagzi et al. (2010)],
- enzyme-based logical circuits [Katz and Privman (2010); Privman et al. (2009)],
- spatially extended crystallization computers for optimization and computational geometry [Adamatzky (2009)],
- Physarum computers [Nakagaki et al. (2000, 2001, 2007)],
- geometrically constrained universal chemical computers [Sielewiesiuk and Górecki (2001); Motoike and Yoshikawa (2003); Górecki et al. (2009); Yoshikawa et al. (2009); Górecki and Górecka (2006,a); Górecki et al. (2003)],
- molecular logical gates and circuits [Stojanovic et al. (2002, 2005); Lederman et al. (2006); Macdonald et al. (2006)].

In contrast, there are thousands of papers on quantum computation and hundreds on membrane computing and artificial immune systems. Such a weak representation of laboratory experiments in the field of unconventional computers could be explained by technical difficulties and costs of prototyping. Chemists and biologists are not usually interested in experimenting with unconventional computers because such activity diverts them from mainstream research in their fields. Computer scientists and mathematicians would like to experiment but are scared of laboratory equipment.

If there was a simple to maintain substrate, which requires minimal equipment to experiment with, then progress in designing novel computing devices would be much more visible.

The chapter is structured as follows. We introduce reaction–diffusion computers, because they are proved to be the most productive experimental implementations of unconventional computers. Then we bring in *P. polycephalum* and provide evidence that plasmodium of *P. polycephalum* can be seen as a reaction–diffusion and excitable medium encapsulated in an elastic membrane. Finally, we overview a brief history of Physarum computing.

1.1   Reaction–diffusion computers

A reaction–diffusion computer is a spatially extended chemical system, which processes information using interacting growing patterns, excitation and diffusive waves [Adamatzky et al. (2005)]. In reaction–diffusion processors, both the data and the results of the computation are encoded as concentration profiles of the reagents. The computation is performed via the spreading and interaction of wave fronts. A great number of chemical laboratory prototypes, designed by De Lacy Costello, are discussed in our previous book [Adamatzky et al. (2005)].

In terms of classical computing architectures, the following characteristics can be attributed to reaction–diffusion computers [Adamatzky (1994, 1996, 2001); Adamatzky et al. (2005)]:

- massive parallelism: there are thousands of elementary processing units, micro-volumes, in a standard chemical vessel;
- local connections: micro-volumes of a non-stirred chemical medium change their states, due to diffusion and reaction, depending on states of, or concentrations of reactants in, their closest neighbors;
- parallel input and output: in chemical reactions with colored prod-

uct the results of the computation can be recorded optically; there is also a range of light-sensitive chemical reactions where data can be inputted by localized illumination;

- fault tolerance: being in liquid phase, chemical reaction–diffusion computers restore their architecture even after a substantial part of the medium is removed; however, the topology and the dynamics of diffusive and particularly phase waves (e.g. excitation waves in a Belousov–Zhabotinsky system) may be affected.

These characteristics of reaction–diffusion chemical computers make them ideally tailored for the implementation of novel and emerging architectures of robotic controllers and embedded processors for smart structures.

Most experimental prototypes of reaction–diffusion computers use a one-to-all type of communication when transmitting information between elementary computing units. This is typical for a Belousov–Zhabotinsky (BZ) medium in excitable mode and precipitating chemical processors.

The first ever experimental chemical processor, presented in [Kuhnert (1986); Kuhnert et al. (1989)], is architectureless and with parallel optical input the illumination gradients of the BZ reactor. Kuhnert's 'BZ-memory' processor [Kuhnert (1986); Kuhnert et al. (1989)] does not employ interaction of propagating excitations but only light sensitivity of the BZ reaction and global switching between excitation and refractory states. The ideas in [Kuhnert (1986); Kuhnert et al. (1989)] are further detailed in [Rambidi (1998); Rambidi et al. (2002)], where a BZ medium was used to extract the contour of a planar shape, detect particular features of the shape and implement negation and enhancement of images and shape restoration.

Other applications of a BZ medium in one-to-all communication mode include approximation of the shortest collision-avoidance path [Agladze et al. (1997)] and a set of all collision-free paths between two planar points [Adamatzky and De Lacy Costello (2002)]. A shortest path is approximated by running excitation waves first from start to destination, then from destination to start and detecting the intersection of wave fronts traveling in opposite directions [Agladze et al. (1997)]. A set of collision-free paths is generated by exciting the medium in all obstacles at once and recording distance fields approximated by traveling wave fronts [Adamatzky and De Lacy Costello (2002)].

A typical precipitating processor operates similarly to BZ processors. Not excitation but diffusive waves propagate. Reactants in the propagating wave fronts precipitate when they react with species in a substrate.

Domains of the substrate covered by diffusive fronts are 'tagged' by precipitate. All precipitating computing devices built so far are based on the single phenomenon: when two or more diffusive wave fronts meet, no precipitate is formed at the meeting loci. That is, the precipitating processor computes a bisector between two planar points. There are several non-trivial designs of precipitating processors which approximate a planar Voronoi diagram [Tolmachev and Adamatzky (1996); De Lacy Costello (2003); De Lacy Costello and Adamatzky (2003); De Lacy Costello et al. (2004a, 2009)] and a skeleton of a planar shape [Adamatzky and Tolmachiev (1997); Adamatzky and De Lacy Costello (2002)]. The recently discovered 'hot ice computer' [Adamatzky (2009)], which exploits crystallization in a supersaturated solution of sodium acetate, allows for 'one-passage' calculation of shortest path.

A one-to-all type of communication is typical for plasmodium of *P. polycephalum* growing on a nutrient substrate (Chap. 3.3).

To implement a one-to-one type of communication in a reaction–diffusion medium, we can use self-localized excitations, wave fragments, traveling in a BZ medium in a sub-excitable mode [Sedina-Nadal et al. (2001)]. The excitation wave fragments in a sub-excitable BZ medium behave like quasi-particles. They exhibit rich dynamics of collisions, including reflection, fission, fusion and annihilation [Adamatzky and De Lacy Costello (2007); Toth et al. (2009)]. Using the wave fragments, we implemented a collision-based computing scheme [Adamatzky (2004)].

Most Physarum machines discussed in the present book use the one-to-one type of communication by plasmodium self-localizations propagating on a non-nutrient substrate.

1.2   Limitations of reaction–diffusion computers

There are problems which reaction–diffusion chemical processors fail to solve. Shortest path and spanning tree problems are typical tasks failed by reaction–diffusion computers. Experimental techniques [Agladze et al. (1997)] and [Adamatzky and De Lacy Costello (2002)] on computation of shortest path in a Belousov–Zhabotinsky medium require significant assistance from conventional computers. Namely, snapshots of traveling excitations must be recorded at regular intervals and analyzed on a PC. Reaction–diffusion computers cannot calculate a spanning tree nor can they implement memory, unless it is ancient coil-type memory in a ring-constrained BZ medium. Geometrical constraining in general does not help to tackle these tasks. Encapsulation looks like the only solution [Adamatzky (2007a)].

A common way to encapsulate a BZ system is to disperse it in a water-in-oil micro-emulsion with surfactant [Epstein (2005)]. BZ reagents are thus enclosed in a mono-layer of anionic surfactant. The BZ droplets are immersed in oil through which intermediate reactants diffuse [Epstein (2005)]. The diffusion of reactants can be seen as a continuous supply of energy to BZ droplets. 'Unlimited' energy supply enables BZ droplets to stay excited for a very long time. This system exhibits a wide range of stationary (similar to Turing structures) and oscillatory patterns. A stationary pattern can be photo-printed on the medium. When imprinted, the patterns are represented by configurations of stationary excitations for as long as reagents are refilled [Kaminaga (2006)].

Experiments with encapsulated chemical media require a certain level of chemical expertise and at least some kind of laboratory equipment. Certainly, it would be quite difficult for you to make a DIY BZ droplet in your kitchen. We went in search of an easy-to-experiment-with analog of an encapsulated reaction–diffusion system and picked plasmodium of *P. polycephalum* as a suitable one.

1.3   *Physarum polycephalum*

Physarum polycephalum belongs to the species of order *Physarales*, subclass *Myxogastromycetidae*, class *Myxomycetes*, division *Myxostelida*. It is commonly known as a true, acellular or multi-headed slime mold. The life cycle of *P. polycephalum* is exciting (Fig. 1.1).

Plasmodium — a 'vegetative' phase – is a single cell with a myriad of diploid nuclei (Fig. 1.1a–l). The plasmodium looks like an amorphous yellowish mass with networks of protoplasmic tubes (Fig. threeexampelsa). The plasmodium behaves and moves as a giant amoeba. It feeds on bacteria, spores and other microbial creatures and micro-particles [Stephenson and Stempen (1984)]. When foraging for its food the plasmodium propagates towards sources of food particles, surrounds them, secretes enzymes and digests the food. Typically, the plasmodium forms a congregation of protoplasm covering the food source. When several sources of nutrients are scattered in the plasmodium's range, the plasmodium forms a network of protoplasmic tubes connecting the masses of protoplasm at the food sources.

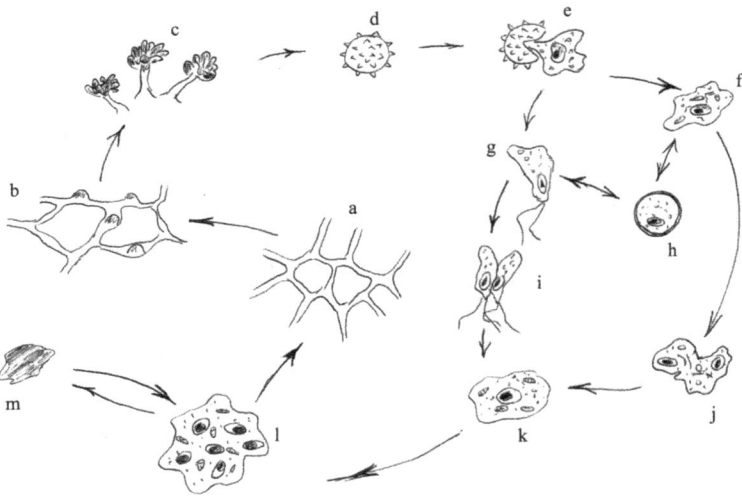

Fig. 1.1 Life-cycle scheme of *Physarum polycephalum*, inspired by [Stephenson and Stempen (1984)]: (a) protoplasmic tree formed by 'vegetative' state — plasmodium, (b) early stage of sporulation, (c) sporangia — fruiting bodies with spores, (d) a single spore, (e) germinated spore — cracked spore with myxamoeba crawling out, (f) myxamoeba, (g) swarm cell, (h) micro-cyst, (i) fusion of two swarm cells, (j) fusion of two myxamoebas, (k) zygote, (l) plasmodium, (m) sclerotium.

When plasmodium is deprived of water and/or nutrients and cannot migrate to better places, the plasmodium goes into 'hibernation' mode and forms a hardened mass called sclerotium (Figs. 1.1m and threeexampelsb). Sclerotia survive a range of very harsh conditions, including high temperatures of up to 70–80°C [Blackwell et al. (1984)]. A sclerotium of *P. polycephalum* consists of "crustose deposit containing nucleated spherules of cytoplasm enclosed within a honeycomb-like matrix of organic walls" [Chet and Henis (1975); Andreson (2007)]. When moistened, the sclerotium gradually returns to the state of plasmodium (Fig. 1.1l). Myxamoebas can live as they are for a long time. In the presence of water a myxamoeba is transformed into a swarm cell with two flagellas (Fig. 1.1g). Swarm cells can swim. Myxamoebas and swarm cells can reproduce asexually, by simple division. During changes of environment from good to bad, myxamoebas and swarm cells can form spheroidal micro-cysts (Fig. 1.1h) with cellulose walls.

(a)   (b)

(c)

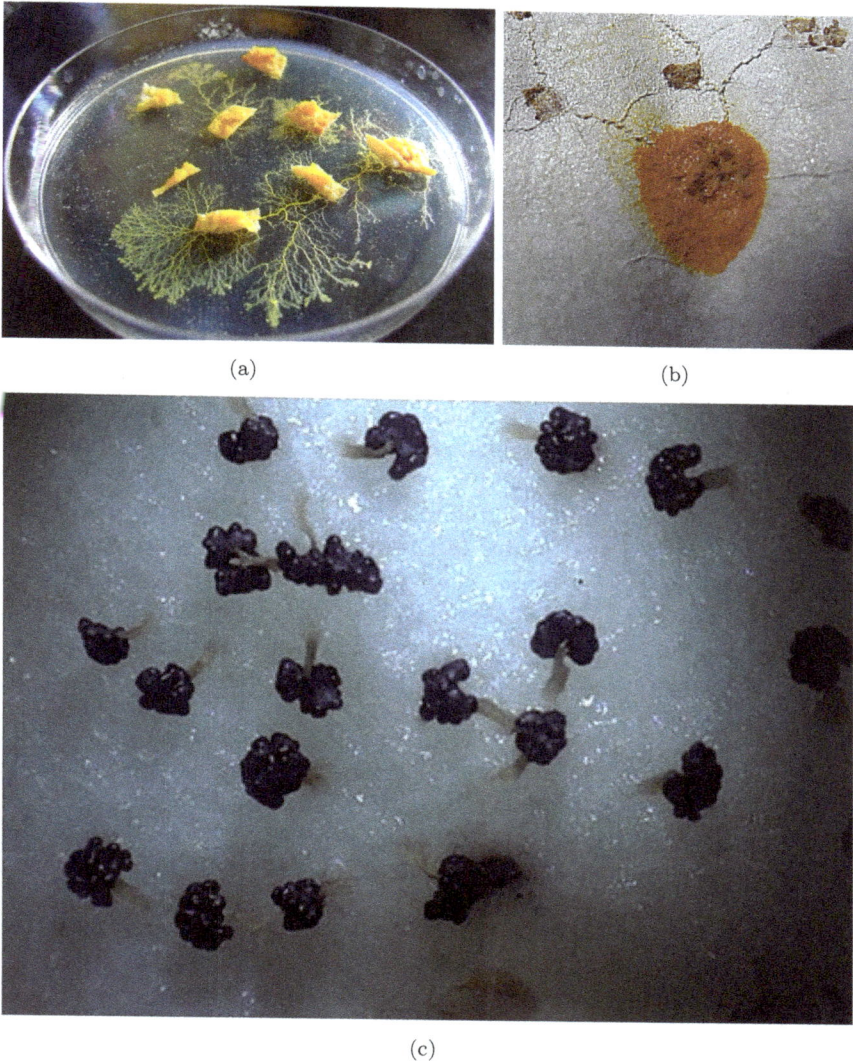

Fig. 1.2   Examples of life forms of *P. polycephalum*: (a) plasmodium, (b) sclerotium, dark-yellowish mass, (c) sporangia.

When exposed to bright light and starved, the plasmodium switches to fructification phase. It grows sporangia (Figs. 1.1b and c and threeexampelsc), finger-like globose enclosures of membrane filled with spiny spores.

When a spore (Fig. 1.1d) gets into a favorable environment, it cracks and releases a single-cell myxamoeba (Fig. 1.1e). When enough myxamoebas or swarm cells are present in the volume, they begin sexual reproduction (Fig. 1.1i and j) and form a zygote (Fig. 1.1k). The zygote divides mitotically and forms a multi-nuclear single cell — the plasmodium (Fig. 1.1l).

1.4   Physarum as encapsulated reaction–diffusion computer

The plasmodium is a network of biochemical oscillators [Matsumoto et al. (1988); Nakagaki et al (1999)]. Waves of excitation or contraction originate from several sources, e.g. induced by external stimuli and perturbations. The waves travel along the plasmodium and interact one with another in collisions. The oscillatory cytoplasm of the plasmodium is a spatially extended nonlinear excitable medium.

Growing and feeding plasmodium exhibits characteristic rhythmic contractions with articulated sources. The contraction waves are associated with waves of electrical potential change. The waves observed in plasmodium [Matsumoto et al. (1986, 1988); Yamada et al. (2007)] are similar to the waves found in excitable chemical systems, like a BZ medium.

The following wave phenomena were discovered experimentally [Yamada et al. (2007)]:

- undisturbed propagation of contraction waves inside the cell body,
- collision and annihilation of contraction waves,
- splitting of the waves by inhomogeneity,
- the formation of spiral waves of contraction.

These are closely matching dynamics of pattern propagation in excitable reaction–diffusion chemical systems [Adamatzky et al. (2005)].

The plasmodium's behavior is determined by external stimuli and excitation waves traveling and interacting inside the plasmodium [Nakagaki et al (1999)]. The plasmodium can be considered as a reaction–diffusion [Adamatzky (2007a)] or an excitable [Achenbach and Weisenseel (1981)] medium encapsulated in an elastic growing membrane. Thus, the plasmodium joins the *Kunstkammer* collection of natural computing substrates additional to existing reaction–diffusion chemical computers [Adamatzky et al. (2005)].

We complete the section with two examples of wave-based 'decision' making in plasmodium.

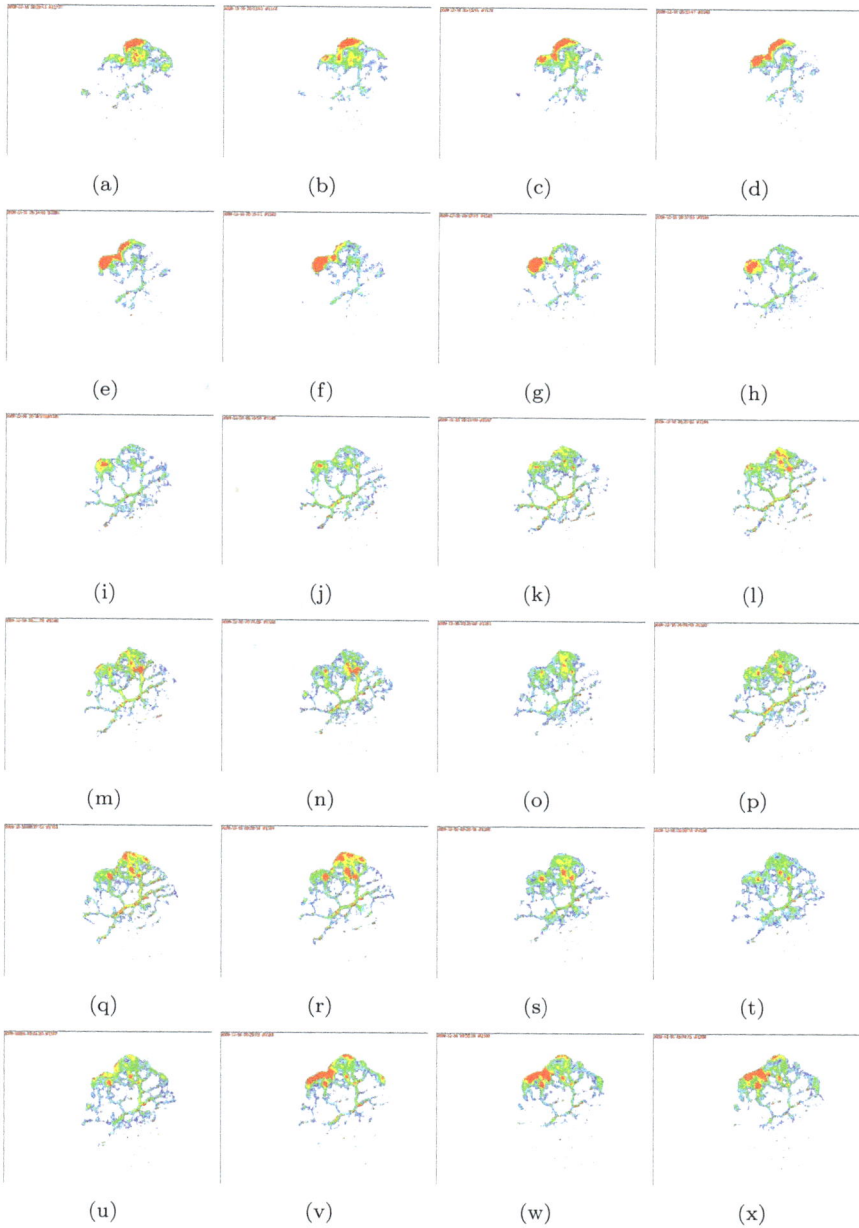

Fig. 1.3　Colorized snapshots of tiny plasmodium blob propagating on a nutrient agar gel. ×40 magnification. One frame per minute recording.

Control of the plasmodium's mobility via propagating waves is demonstrated in Fig. 1.3. We made a small incision in a wall of a thick protoplasmic tube. A drop (about 0.1-mm radius) of cytoplasm popped out. We picked up this drop and transferred it to a thin plate of oatmeal agar. The plate is illuminated from the south-eastern corner, so the gradient of illumination decreases towards the north-western corner of the plate. We record the behavior of the drop with ×40 magnification; the time interval recording is one frame per minute. To detect domains of plasmodium with increased concentration of cytoplasm, we colorize the original images (the values of red and green components of the original image pixels are discretized in five intervals). Pixels with the highest value of red or green components are assigned red color, then yellow, green, blue-green and blue. Pixels with red or green components below a certain threshold are assigned white color. We see that waves are generated on actively propagating parts of the plasmodium blob. The waves then travel inside the plasmodium and along its protoplasmic network, which trails the propagating wave front. The western part of the plasmodium wave front in Fig. 1.3 acts as a generator of traveling waves. Thus, the plasmodium gradually turns west to adjust its motion along the north-western axis Fig. 1.3.

In the experiment illustrated in Fig. 1.4, we show how a plasmodium behaves when faced with sources of repellents and attractants at once. A source of light positioned near the south-eastern corner of the agar plate acts as a repellent. An oat flake, seen as a stationary solid red domain in Fig. 1.4, acts as an attractant. We witness that two sources of contractile waves emerge in the plasmodium. One source, the generation of which is visible in Fig. 1.4a–p, is formed in the part of the plasmodium distant from the source of illumination. This wave source leads the plasmodium towards an escape route from the illumination. At the same time, a second source of waves (Fig. 1.4q–x) is generated in the part of the plasmodium close to the oat flake. This wave source leads the plasmodium towards the attractant, the oat flake. During about 30 h of experiment we observed that the plasmodium circles a few times, approaching the oat flake and then going away. Eventually, the 'need' to escape from light prevails and the plasmodium moves north-west.

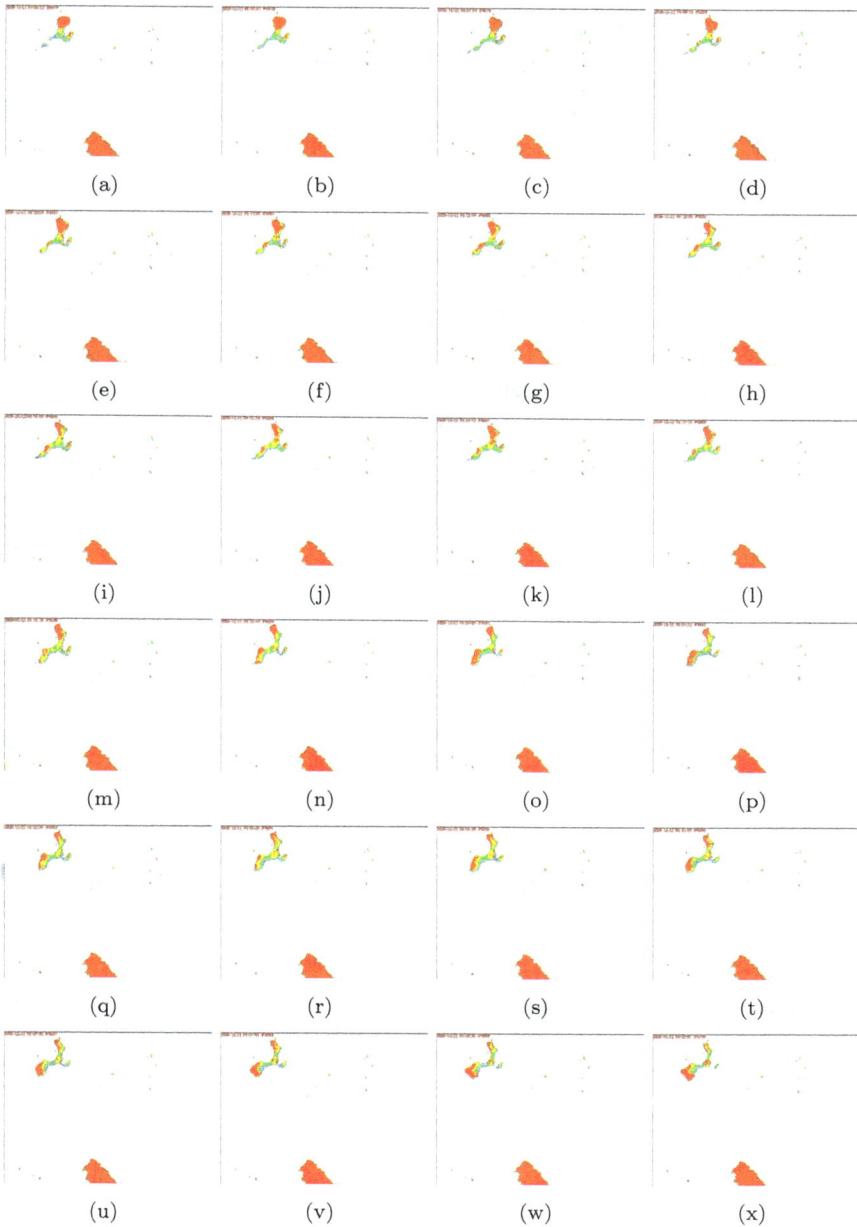

Fig. 1.4 Wave-based decision making by plasmodium. Colorized snapshots of tiny plasmodium blob propagating on a non-nutrient agar gel. ×40 magnification. One frame per minute recording. Stationary red domain at the southern edge of the agar plate is an oat flake.

1.5   Dawn of Physarum computing

In 2000 *Nature* published a paper 'Maze-solving by an amoeboid organism' by Toshiyuki Nakagaki, Hiroyasu Yamada and Ágota Tóth [Nakagaki et al. (2000)]. The authors experimentally proved that — when source and destination are represented by sources of food — the plasmodium of *P. polycephalum* approximates a shortest path in the labyrinth by developing a thick protoplasmic tube connecting the source and the destination.

When placed in an environment with distributed sources of nutrients the plasmodium forms a network of protoplasmic tubes connecting the food sources. If we interpret sources of food (e.g. oat flakes) as nodes and protoplasmic tubes as edges, we see that the plasmodium constructs a planar graph on the sources of food. Nakagaki et al. [Nakagaki et al. (2001, 2007)] showed that the topology of the plasmodium's protoplasmic network optimizes the plasmodium's harvesting on the scattered sources of nutrients and makes more efficient the flow and transport of intracellular components.

In the last 10 years, experimental laboratory prototypes of Physarum computers were designed to

- compute a shortest path [Nakagaki et al. (2000, 2001, 2007); Shirakawa and Gunji (2009)],
- implement logical gates [Tsuda et al. (2004)],
- control a robot; plasmodium acts as an on-board controller for robot photo-avoidance [Tsuda et al. (2007)],
- approximate proximity graphs [Adamatzky (2008)],
- compute Voronoi diagrams [Shirakawa et al. (2009); Shirakawa and Gunji (2009)],
- construct spanning trees [Adamatzky (2007,a)],
- implement storage-modification machines [Adamatzky (2007)],
- solve resource-consuming computational problems [Aono and Gunji (2001, 2004); Tsuda et al. (2004); Nakagaki et al. (2001); Tsuda et al. (2006)],
- approximate Delaunay triangulation [Shirakawa et al. (2009)],
- implement primitive memory [Saigusa et el. (2008)],
- implement spatial logic and process algebra [Schumann and Adamatzky (2009)].

The plasmodium functions as a parallel amorphous computer with parallel inputs and parallel outputs. Data are represented by spatial configurations of sources of nutrients. A program of computation is coded via configurations of repellents and attractants. Results of the computation are presented by the configuration of the protoplasmic network and the localization of the plasmodium. In the book we demonstrate that plasmodium of *P. polycephalum* is a parallel computing substrate complementary to, and sometimes more efficient than, existing massively parallel reaction–diffusion chemical processors [Adamatzky et al. (2005)].

Chapter 2

Experimenting with Physarum

2.1 Where to get plasmodium of *P. polycephalum*

If you do not have immediate friends cultivating Physarum[1], you have three options.

First, you can buy it for $10 from Carolina Biological Supply[2].

Second, go to a nearby forest (check your garden beforehand — maybe Physarum is already there) and try to find a sample. Have a look in [CalPhotos (2009); Discover Life (2009)] to know what the plasmodium of *P. polycephalum* looks like in natural conditions. Read [Stephenson and Stempen (1984); Ing (1999)] helpful hints on collection of samples. There is one book on *Myxomycetes* you can read for free, thanks to the Gutenberg project [Morgan (1893)].

The third option would be to search on Google for those who do experiment with *P. polycephalum*, explain to them why you need it and hopefully they will be kind enough to post a sample to you. More likely you will receive a dried sclerotium on a piece of paper. Place the piece with sclerotium somewhere, wet generously and in one or two days the plasmodium will wake up.

2.2 Physarum farms

You can cultivate plasmodium on any porous substrate; it is better to use disposable ones, e.g. paper kitchen towels, toilet paper, paper napkins/serviettes. Any container is fine. I am using plastic lunch boxes and

---

[1]I was lucky to get a sclerotium of *P. polycephalum* directly from Dr. Soichiro Tsuda. He even supplied instructions. What a gentleman.

[2]http://www.carolina.com

Fig. 2.1   Excessive growth of Physarum.

empty ice-cream boxes. Moisten the substrate every couple of days, or when it becomes dry; and replant to fresh substrate every week or two. An expert hint from [Ing (1999)]: if it is too smelly — it is too wet, if it is too

(a)

(b)

Fig. 2.2 'Green' disposal of Physarum. (a) One shrimp tastes synticium, soon the bacchanalia starts (b).

Fig. 2.3 Petri dishes with plasmodium are stored at room temperature in a dark cupboard, a cabinet or a locker.

hairy (overgrown by bacteria or fungi) — it is too dry. The plasmodium is usually fed with oat flakes. However, you can try a variety of foods and see what will happen. With some patience and a bit of luck you establish a routine to keep a sustainable growth of Physarum colonies, but often this will go beyond reasonable needs in your production (Fig. 2.1). How you dispose of the excess is your moral choice. I never kill Physarum pointlessly. I feed excess of Physarum to my fresh-water friends who, particularly shrimps, just adore it (Fig. 2.2).

2.3 Dishes and scanners

For experiments, we use glass or plastic round (90 mm and 25 mm) and square (12 mm) Petri dishes. The substrates are wet filter paper, nutrient-free 2% agar or oat meal 2% agar. (Dis)advantages of substrates are discussed in Sect. 2.5. The Petri dishes with Physarum are stored in a dark place at room temperature (Fig. 2.3). They are only exposed to light during observation and recording of images.

The results of computation by a protoplasmic network are built by plasmodium. Literally any modern photo-recording device can be used to make snapshots or videos of plasmodium developments. At various stages of my life with plasmodium I used a PS2 EyeToy, a camera built into a Sony Ericsson 800i mobile phone, a Logitech Web Cam Pro and a FujiPix S6500 digital camera. They all worked fine.

I found that scanning is better than photographing. Most images of the experiments in this book are produced on a HP Scanjet or Epson Perfection scanner. Magnified images are obtained using a Digital Blue QX-5 computer microscope, outdated but robust and functional.

2.4 Data input with food

We represent data by the spatial configuration of oat flakes. In many Physarum-computing papers you can see the sentence "Physarum does not eat oat flakes but only bacteria growing on the oat flakes.". This is a myth. Possibly one person wrote this a long time without due consideration and others just started to repeat it without checking. Plasmodium consumes all kinds of tiny particles, including pieces of oat flakes (see experimental evidence in Chap. 13.9).

The importance of oat flakes is exaggerated. To test the claim, we selected the most common stuff from a UK household product basket (Fig. 2.4) and fed it to Physarum in various combinations (Figs. 2.5, 2.6, 2.7 and 2.8). Most of the products tested are proved to act as chemoattractants for the plasmodium. Let us consider a few examples[3].

When faced with a choice of apple and honey, the plasmodium does not hesitate and propagates towards the apple and the honey simultaneously (Fig. 2.5a). Honey and sugar powder produce better gradients of attractants than large crystals of sugar (Fig. 2.5b).

In the experiment with carbohydrates, the plasmodium preferred Deli wrapper. Initially the plasmodium develops two processes, passing somewhere along bisectors separating wrapper and pasta in the north-west, and bread and pasta in the south-east (Fig. 2.6a). Then the plasmodium 'discovers' the piece of Deli wrapper, colonizes it and abandons all previous

---

[3]Food preferences discussed here are not statistically proved. We did not aim to discover what Physarum likes most of all. This may be a good exercise for a curious reader.

Fig. 2.4   Products tested with Physarum: Mission Deli Multigrain Wraps (Mission Foods, UK), Kellogg's Crunchy Nut Clusters, Kellogg's Crunchy Nut Corn Flakes, Asda Pork Luncheon Meat, Asda Sliced Pork and Egg Roll, Capilano Natural Australian Honey (Capilano Honey Ltd), Asda Porridge Oats and Warburton's Farmhouse Soft White Bread (Warburton Ltd).

tubes (Fig. 2.6b). The piece of wrapper becomes the advance post for further exploration of the substrate (Fig. 2.6c).

Plasmodium does prefer sugary flakes to oat meal flakes (Fig. 2.7); glucose–fructose syrup covering the flakes plays a crucial role. In our 'egg vs. meat' experiments, the plasmodium showed some preference to egg white (Fig. 2.8a and b); luncheon meat was the plasmodium's second choice (Fig. 2.8c).

2.5   Substrates

It is a pleasure to experiment with plasmodium. Physarum is a pretty tough creature. It can propagate and survive, at least for a short time, on a variety of substrates. Plasmodium successfully spans bare glass and plastic (Fig. 2.9a–c), and even arrays of naked electrodes (Fig. 2.9d).

The plasmodium does not feel too comfortable on aluminum foil (Fig. 2.10). However, even in such an environment it tries to forage for food (Fig. 2.10a and b) and partially spans sources of nutrients (Fig. 2.10c).

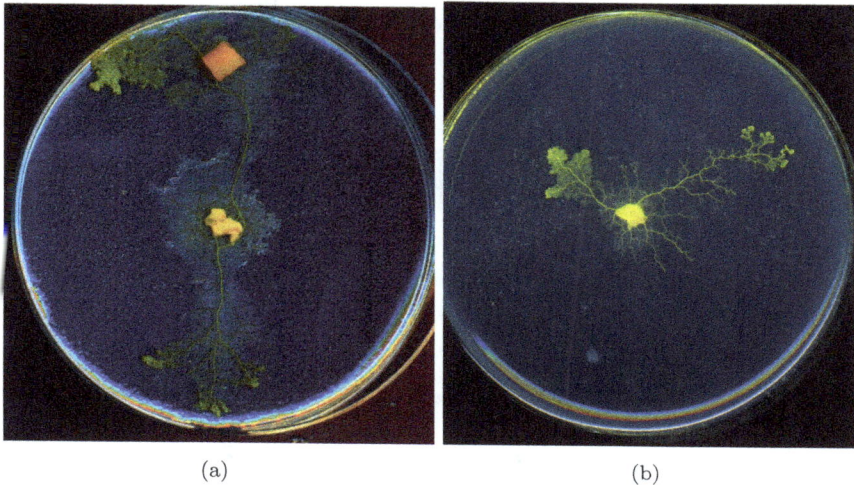

(a)   (b)

Fig. 2.5 Sugary stuff: (a) apple (12 o'clock) vs. honey (6 o'clock) and (b) honey (10 o'clock), sugar powder (1 o'clock), sugar (6 o'clock).

Being curious as to how the plasmodium acts in an environment with high temperature and strong vibrations, we fixed a Petri dish with Physarum inside a MK5 Ford Transit Van engine compartment (Fig. 2.11a). We did not measure the level of noise nor the frequency of vibration; everyone who drove or has ridden in a Transit Van knows very well how noisy they are. Dynamics of temperature changes[4] inside the engine compartment during two days are shown in Fig. 2.11b. The lowest temperature of 13°C was at night, the highest reached 45°C 15–20 min after start of driving.

The plasmodium was inoculated at the center of the Petri dish on top of wet filter paper. Temperature and vibration depress Physarum, slow down its growth and suppress its foraging activities. In this particular experiment, it took the plasmodium almost one day to propagate from its initial position to the eastern oat flake (Fig. 2.11c). In the next 22 h the plasmodium attempted to link the eastern oat flake with the north-north-western group of oat flakes (Fig. 2.11d). The plasmodium formed sclerotium after that.

Silicone gum[5] may provide a protection for plasmodium computers func-

---

[4]The temperature is recorded using a Lascar USB temperature data logger EL-USB-1; the logger was positioned near the Petri dish.

[5]Octamethylcyclotetrasiloxane, trade name Silastic 4-2735 Silicone Gum, Dow Corning

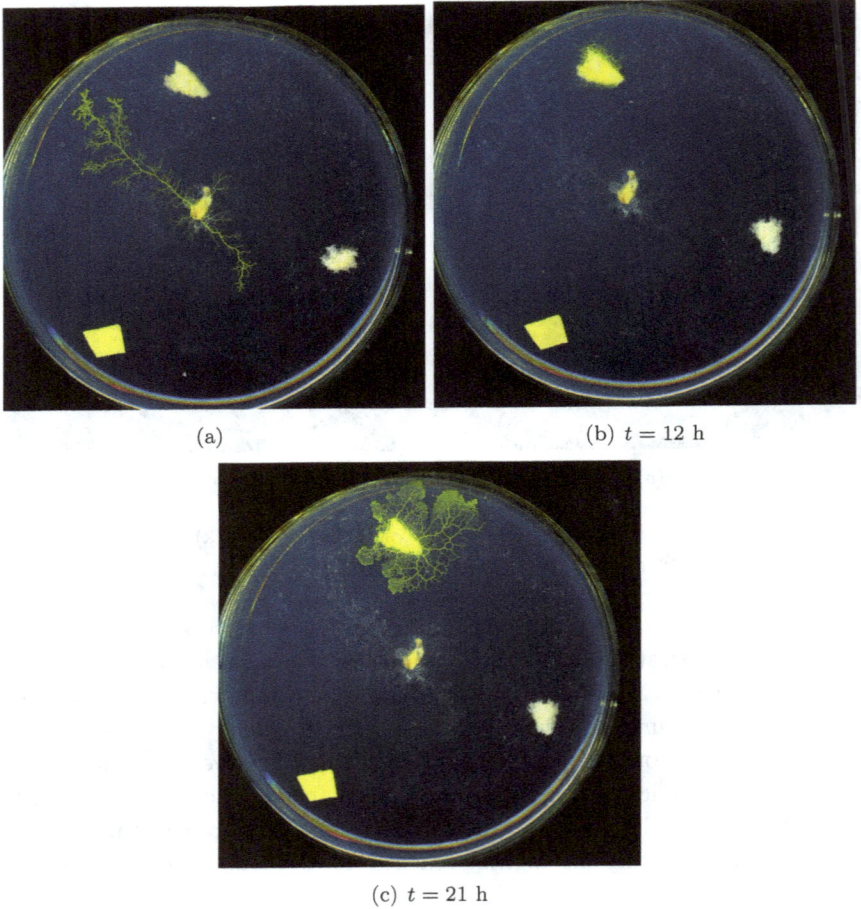

(a)   (b) $t = 12$ h

(c) $t = 21$ h

Fig. 2.6   Carbohydrates: (a) bread (3 o'clock), (b) pasta (7 o'clock), (c) Deli wrapper (12 o'clock).

tioning in a harsh environment. It is possible to prepare a flexible and relatively stable silicone gel plate with living protoplasmic tubes inside (Fig. 2.12). When plasmodium propagates in the silicone gum, the plasmodium's active zone, or growing tip, stays on the surface of the gel layer, while a protoplasmic tube unreeled from the tip sinks inside the gel.

In summary, Physarum can live for a short time on a very wide range of substrates and in a very wide range of environmental conditions. However,

S.A., B-7180 Seneffe, Belgium.

(a)

(b) $t = 8$ h

(c) $t = 20$ h

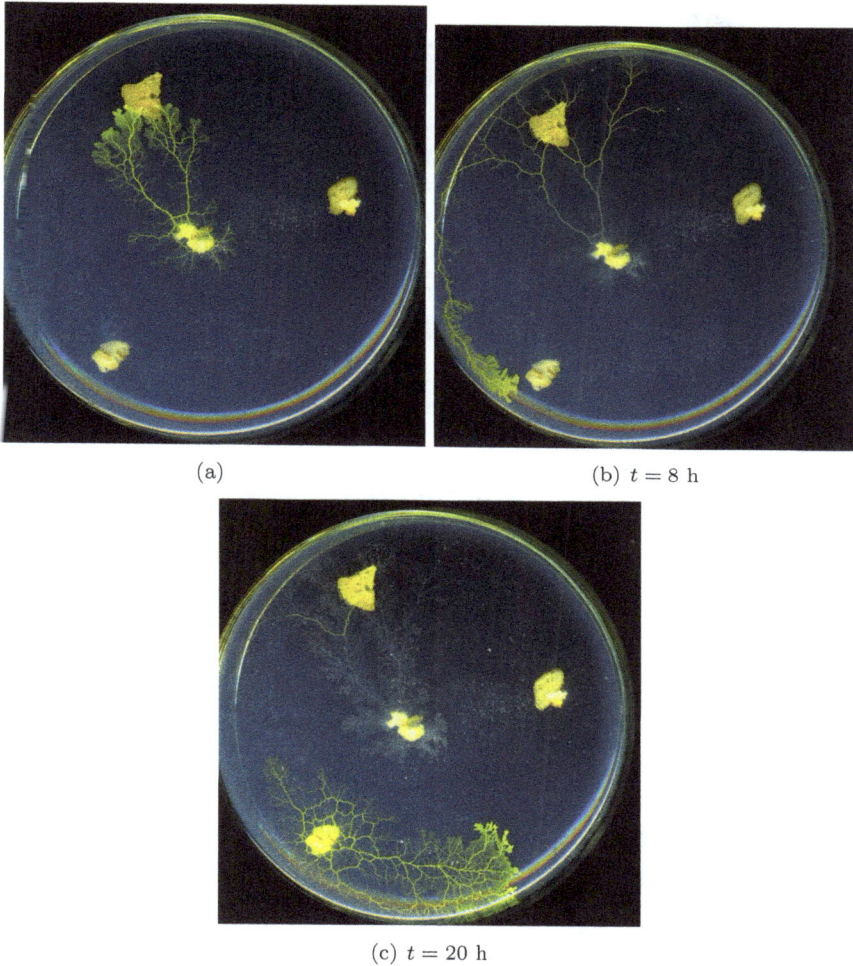

Fig. 2.7 Flaky–crunchy: Crunchy Nut Corn Flake (11 o'clock), Crunchy Nut Cluster (3 o'clock), oat meal flake (7 o'clock).

it is always better to cultivate the plasmodium on agar gel or humid filter paper during prolonged experiments. In some cases it is enough to make at least some domains of the experimental space Physarum friendly, as shown in experiments with agar gel tiles in Fig. 2.13.

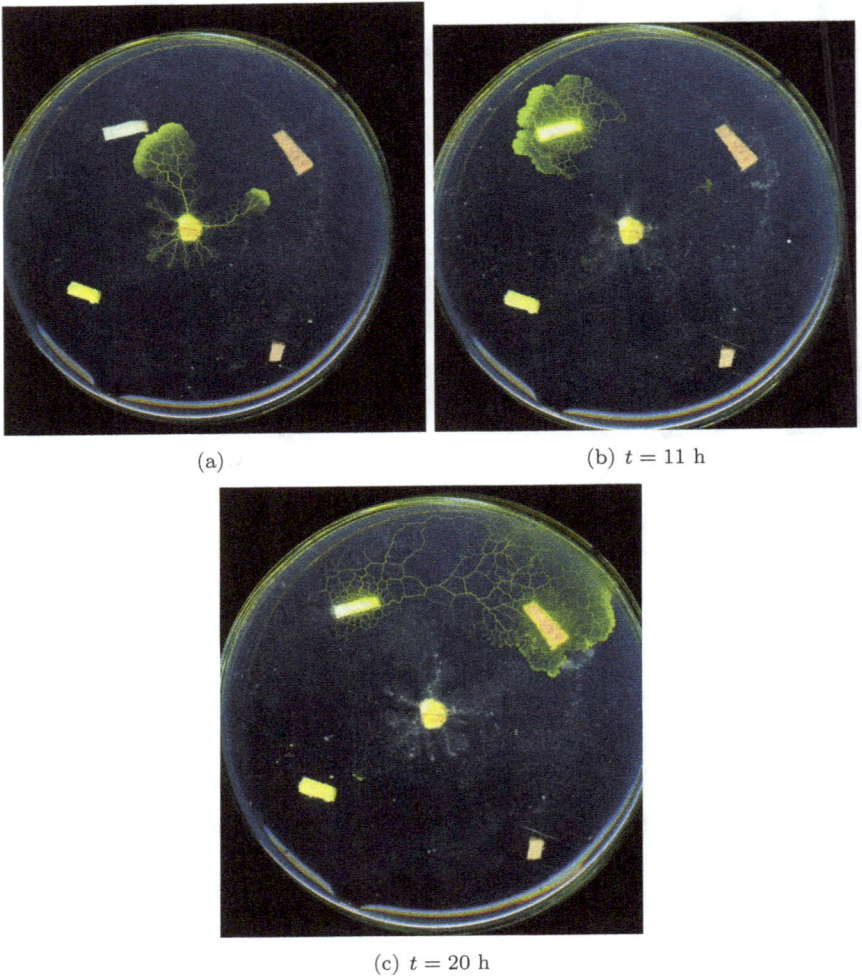

(a)

(b) $t = 11$ h

(c) $t = 20$ h

Fig. 2.8   Meat and boiled egg: egg white (11 o'clock), luncheon meat (1 o'clock), pork roll (5 o'clock), egg yolk (8 o'clock).

2.6   Nutrient-rich vs. non-nutrient substrates

Most experiments discussed in the book are undertaken with Physarum growing on a nutrient-free, or non-nutrient, substrate. We use 2% agar gel (Select Agar, Sigma Aldrich) as a non-nutrient substrate and 2% corn meal agar gel (Corn Meal Agar, Fluka Analytical) as nutrient-rich, or nutrient,

Fig. 2.9   Plasmodium grows on a variety of unusual substrates: (a) empty glass, (b) bare plastic Petri dish, (c) toy car, (d) multi-electrode array.

substrate. We found — and use this all through the book — the following fact.

Finding 1. *The plasmodium propagates as a typical diffusion/excitation wave on a nutrient substrate. The plasmodium propagates as a localization, similar to localized wave fragments and dissipative solitons, on a non-nutrient substrate.*

We show the difference between nutrient and non-nutrient substrates in the two following examples. In Fig. 2.14, two pieces of plasmodium are inoculated at the boundary between non-nutrient and nutrient agar plates (Fig. 2.14a). The plasmodia definitely prefer nutrient agar. Waves of the plasmodium growth are clearly visible on the nutrient plate (Fig. 2.14b). This experiment shows that — when given a choice – Physarum grows only on a nutrient agar.

(a) $t = 0$ h   (b) $t = 4$ h

(c) $t = 10$ h

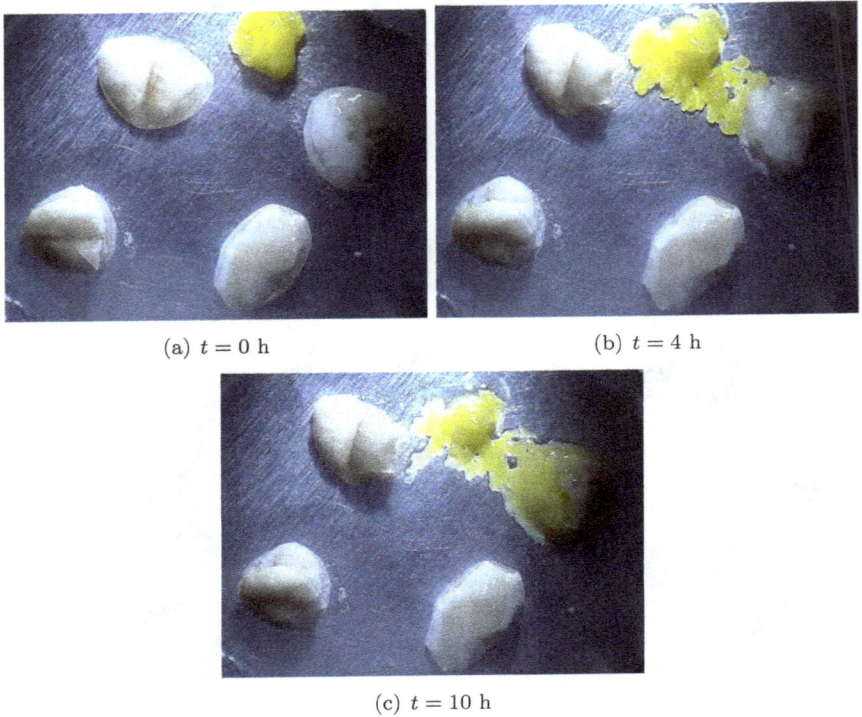

Fig. 2.10   Plasmodium grows on aluminum foil. $\times 10$ magnification.

In the experiments illustrated in Fig. 2.15, we did not provide Physarum with a choice.

In a first experiment, we filled a space around a non-nutrient agar disk with nutrient gel (Fig. 2.15a and b) and inoculated the plasmodium on a non-nutrient disk. The plasmodium propagates as a tree-like structure on the non-nutrient disk (Fig. 2.15a). It continues as a wave-like pattern on the nutrient agar (Fig. 2.15b).

The relation between types of substrate is reversed in a second experiment (Fig. 2.15c and d). The plasmodium is inoculated on a nutrient agar disk surrounded by a non-nutrient agar disk. The growing pattern is transformed from wave-like (Fig. 2.15c) to tree-like (Fig. 2.15d) when the plasmodium passes from the nutrient agar disk to its non-nutrient surroundings.

(a)

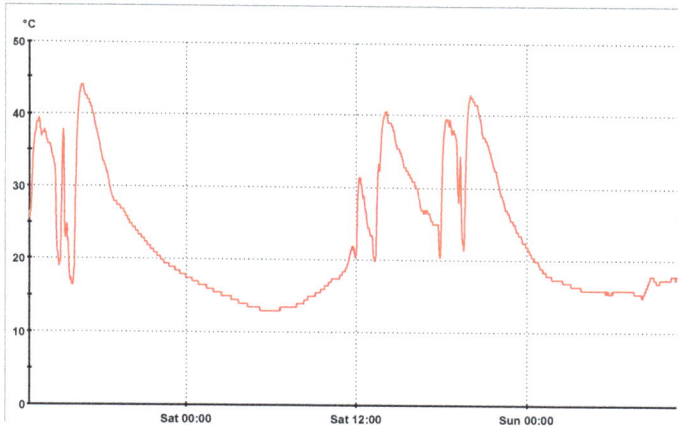

(b)

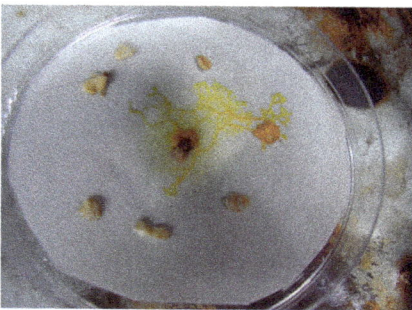

(c)

(d)

Fig. 2.11 Physarum's development in hot environment with strong vibrations: (a) position of Petri dish with Physarum in engine compartment, (b) temperature inside the engine compartment during two days since inoculation of plasmodium, (c) development of plasmodium 21 h after inoculation, (d) plasmodium 43 h after inoculation.

Fig. 2.12   Physarum propagating on silicone gum. Growing tips stay on the surface of the gel, while protoplasmic tubes sink.

A nutrient substrate (corn meal agar gel) acts as a spatially extended energy supply for plasmodium. This is why growing plasmodium behaves like an excitable, or auto, wave. The plasmodium even shows — typical for an excitable medium — the phenomenon of rotating tips of spiral waves (Fig. 2.16).

When plasmodium is inoculated, on an oat flake it is attached to, on a non-nutrient substrate, it has a limited amount of energy. This energy limitation makes the plasmodium propagate as a self-localized pattern. This traveling localization leaves just one protoplasmic tube behind (Fig. 2.17). As soon as the plasmodium reaches a 'fresh' source of nutrients, an intact oat flake, the plasmodium gets an additional supply of energy and expands its propagating pattern (Fig. 2.17).

Finding 2. *S*ources of nutrients scattered on a non-nutrient agar act as natural amplifiers of plasmodium self-localizations.

2.7   Sensing

Any source of food placed in a plasmodium's vicinity emits certain types of chemicals. The chemicals diffuse in the substrate. The plasmodium propagates along these gradients of chemo-attractants towards a source of nutrients.

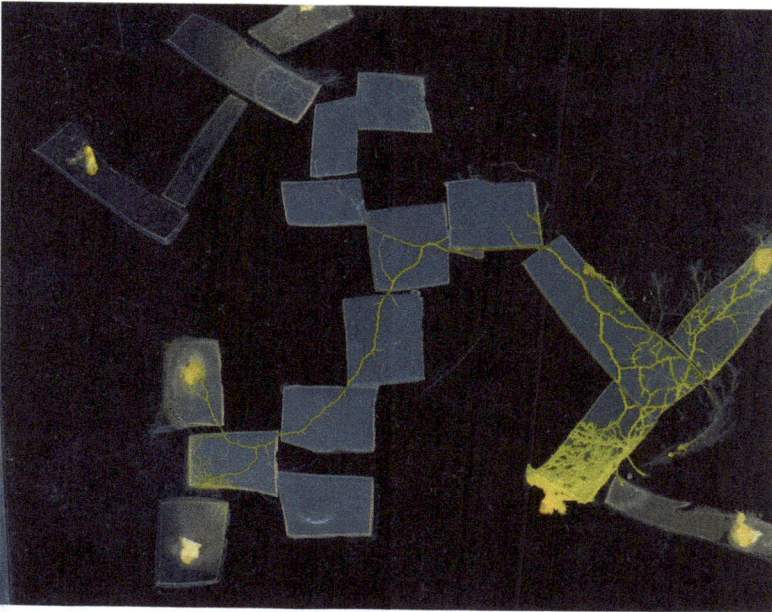

Fig. 2.13   Physarum propagating on a set of agar gel tiles.

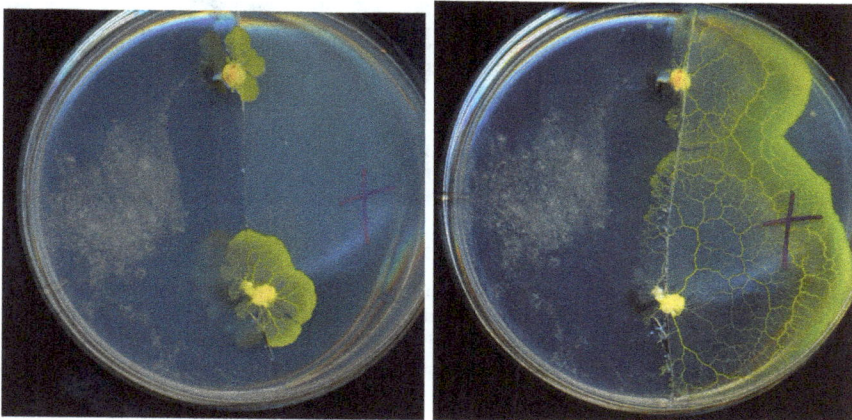

(a) $t=0$ h   (b) $t=18$ h

Fig. 2.14   Growth of plasmodium on two joined plates of agar gel: (a) non-nutrient agar, (b) nutrient agar. The plate of nutrient agar is marked with cross.

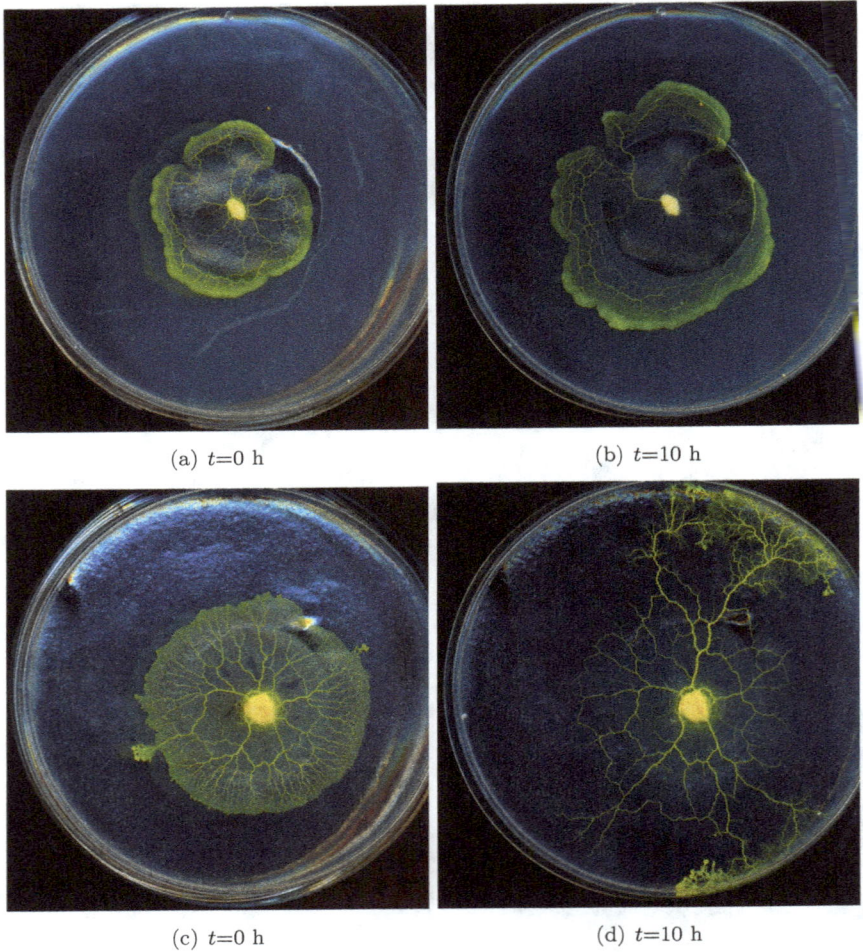

(a) $t=0$ h   (b) $t=10$ h

(c) $t=0$ h   (d) $t=10$ h

Fig. 2.15   Growth of plasmodium on a combination of non-nutrient and nutrient agar shapes: (a) and (b) disk of agar surrounded by oat meal agar, (c) and (d) disk of oat meal agar surrounded by agar; the plasmodium does not have a choice but, to continue growing on the substrate provided, it becomes choosy when it can; see Fig. 2.14.

If no chemo-attractants are present in the plasmodium's environment, the plasmodium exhibits exploratory activity. It propagates in a disorganized manner and branches often. This type of scouting activity is reflected in random-like morphology of protoplasmic tubes and bushy trees

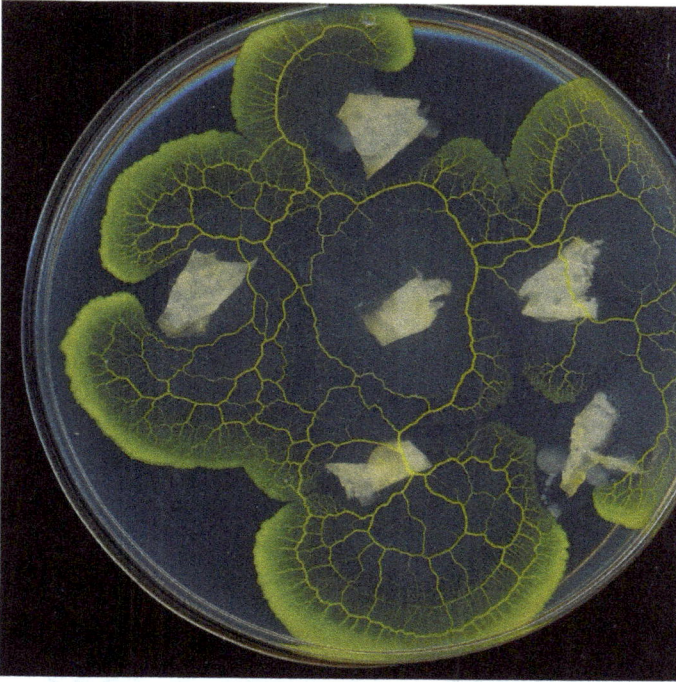

Fig. 2.16   Rotating tips of plasmodium waves. The plasmodium propagates on a nutrient substrate, oat meal agar. Several pieces of plasmodium, on filter paper, are placed on the agar gel.

(Fig. 2.18a). When a source of nutrients is placed in the experimental arena — and if the current state of the substrate permits speedy diffusion of chemo-attractants — the plasmodium quickly detects the direction towards the source of attractants and steadily propagates towards the source (Fig. 2.18b and c).

How reliable is plasmodium in locating sources of nutrients? It is absolutely reliable when nutrients are nearby. If a group of oat flakes is located 2 cm apart of a plasmodium the plasmodium propagates towards the flakes along an almost straight trajectory (Fig. 2.19c). The plasmodium is less confident when sources of chemo-attractants are far from it (Fig. 2.19b).

Let us compare two situations: no source of nutrients (Fig. 2.19a) and one source of nutrients (Fig. 2.19b). In the absence of nutrients the plasmodium's trajectories fill the Petri dish in a disorganized fashion. Indeed, by scouting its substrate, the plasmodium eventually reaches a distant pole

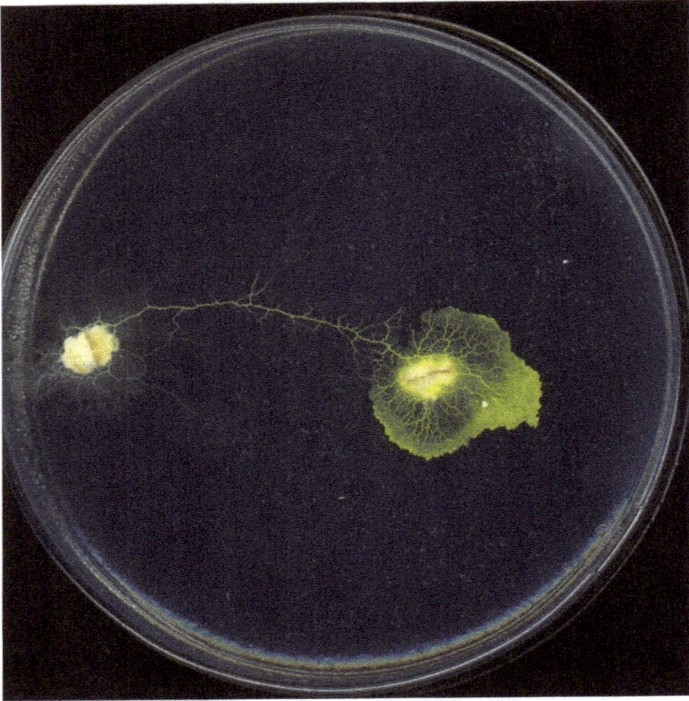

Fig. 2.17   Amplification of plasmodium traveling localization by geometrically restricted source of nutrients, an oat flake. The plasmodium propagates on a non-nutrient substrate, agar gel. The plasmodium is inoculated on an oat flake in the western part of the Petri dish. It propagates and colonizes an oat flake in the eastern part of the dish.

of the Petri dish. It often makes more than 90 degree turns.

When a source of chemo-attractants is present the plasmodium propagates towards the source along natural boundaries, walls of the Petri dish or straight forward. In nearly 40% of experiments the plasmodium chooses a direct route through the central part of the Petri dish. Clusters of the plasmodium trajectories are clearly visible in Fig. 2.19b.

To check the plasmodium's ability of detecting chemo-attractants diffusing in the air, we filled a Petri dish with non-nutrient agar and inoculated the plasmodium on the agar. We then fixed a few oat flakes, using Blu-Tack (Bostik Inc.[6]), to the inside of the plastic lid of the Petri dish, and turned the Petri dish upside down. In some experiments the plasmodium managed to propagate towards the projection of oat flakes from the dish's

---

[6]http://www.bostik.co.uk/

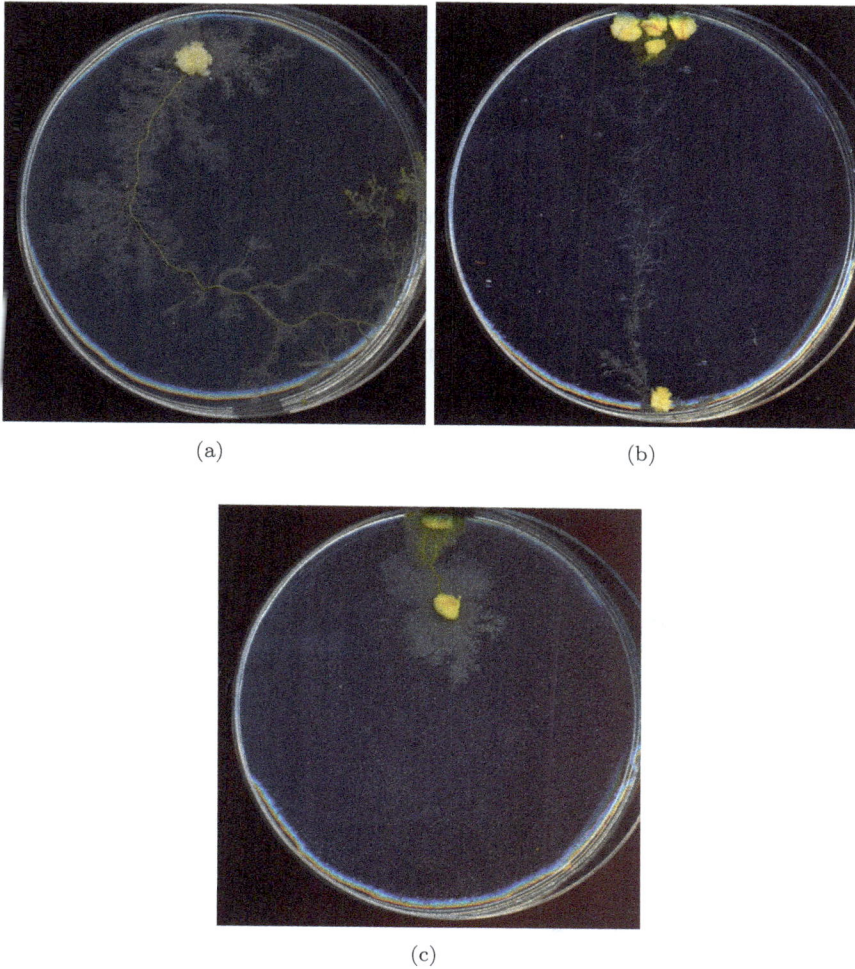

(a)   (b)

(c)

Fig. 2.18   Examples of plasmodium-propagation trajectories: (a) no target food source, (b) target food source is positioned 8 cm apart plasmodium, (c) target food source is positioned 2 cm apart plasmodium.

lid, position the active zone above the source of nutrients and descend down on the target group of oat flakes (Fig. 2.20).

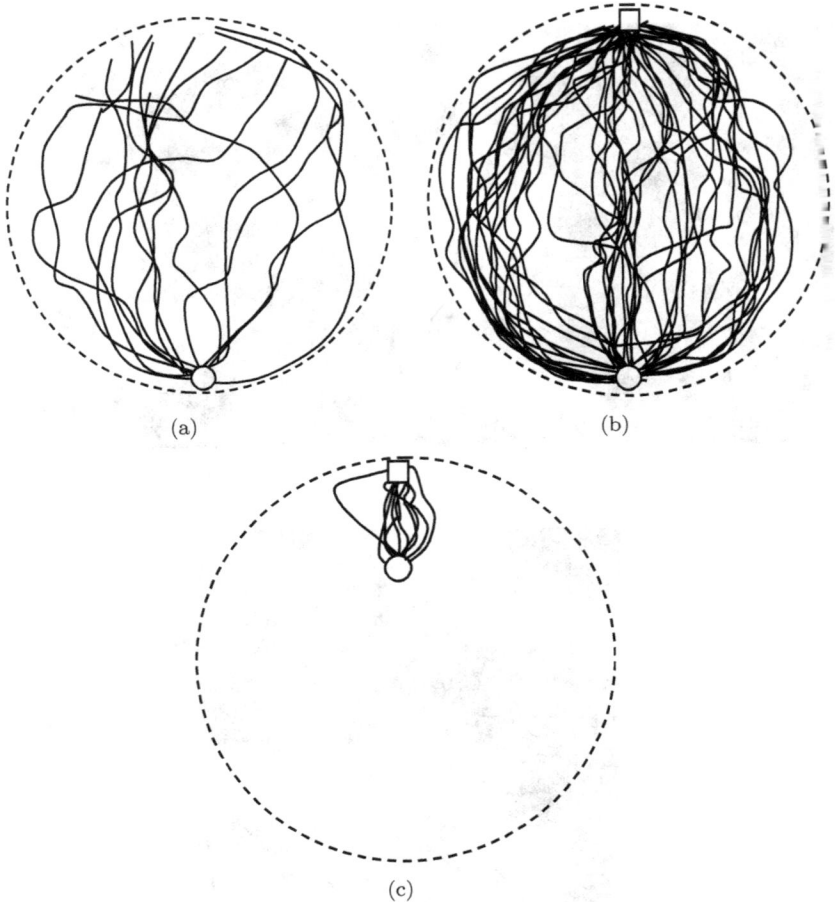

Fig. 2.19 Trajectories of plasmodium propagation on a non-nutrient agar. Initial position of plasmodium is marked by circle. Target oat flakes (if any) are marked by rectangle: (a) no target food sources are present, results of 15 experiments; (b) target food source is positioned opposite, to the site of initial inoculation of plasmodium, part of the Petri dish (approx. 8-cm distance), trajectories of 50 experiments; (c) target food source is positioned 2 cm from plasmodium, 10 experiments.

2.8   Modeling plasmodium

A sophisticated and realistic computer model of plasmodium behavior was developed by Jones [Jones (2010, 2008a, 2009a)]. The model employs a collective of virtual mobile chemo-tactic particles. The particles are con-

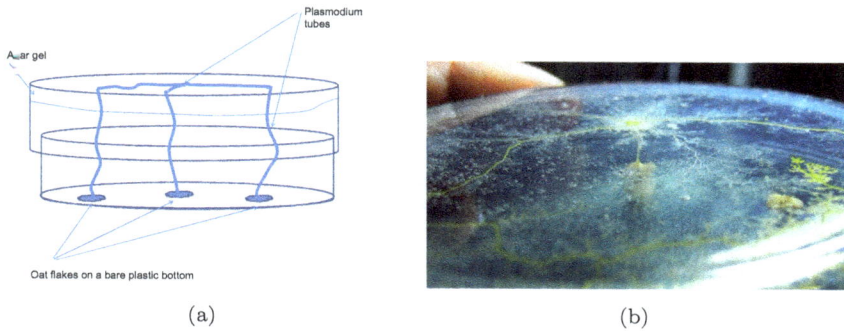

(a)   (b)

Fig. 2.20   Plasmodium, placed on the inside of the Petri dish's lid, detects chemo-attractants in the air, locates the physical position of the source and descends down to the source of nutrients: (a) scheme, (b) photograph of experiment, protoplasmic tube connecting lid and dish is in the center of the Petri dish.

trolled by diffusive gradients of attracting and repelling fields. The particles can generate chemical fields themselves. The collectives of particles merge into propagating streams which morphologically resemble protoplasmic networks of *P. polycephalum*. The Jones model is a key simulation tool to verify experiments on formation of a proximity graph by growing plasmodium [Jones   (2009)], restructuring of protoplasmic networks [Adamatzky and Jones   (2009)] and testing plasmodium's road-building potential [Adamatzky and Jones (2009a)].

In the book we use a 'more physical' model of Physarum propagation: in Chap. 4.3 we adapt a two-variable Oregonator model to simulate plasmodium behavior.

2.9   Summary

Plasmodium of *P. polycephalum* is a virtually free substrate for experiments on unconventional computing. You can collect it in your garden or a nearby forest. The main consumables — Petri dishes, agar, filter paper — used in our experiments are easily available in specialized shops and on **eBay**. Digital cameras are nowadays household items. Therefore, you could not make lack of funds an excuse for failing to reproduce experiments discussed in the book. You do not need expensive equipment to make your personal contribution to the exciting field of unconventional computing.

Chapter 3

Physarum solves mazes

To solve a maze, one must build a route from the start to the finish. A maze without loops is equivalent to a tree, rooted in the central chamber of the maze. Solving a maze is equivalent to computing the shortest path on a tree graph. A shortest-path problem is solved in a variety of experimental laboratory prototypes of unconventional computers: reaction–diffusion chemical [Adamatzky et al. (2005)], gas-discharge analog [Reyes et al. (2002)], micro-fluidic network [Fuerstman et al. (2003)], chemotactic droplet [Lagzi et al. (2010)] and spatially extended crystallization [Adamatzky (2009)] systems. The shortest routes between the nest and sources of food supply are ordinarily built by ants in their foraging activities [Goss et al. (1989); Beckers et al. (1989)]. Analogically, we can assume that a shortest path is approximated by plasmodium of *P. polycephalum*. This is indeed true and was experimentally proved in [Nakagaki et al. (2000, 2001)].

3.1   Multiple-site start

In [Nakagaki et al. (2000, 2001)], walls of a maze are represented by a plastic film template placed on top of an agar substrate. Pieces of plasmodium on a substrate are scattered in the maze. Plasmodia propagate from original sites of inoculation along channels, avoiding the plastic film. Plasmodia originating from different sites merge into a single cell. Pieces of nutrients are placed then in the start and the exit of the maze. The plasmodium forms a pronounced protoplasmic tube, connecting the start and the exit. This thick tube represents a path from the start to the exit of the maze.

Such an approach is a little inefficient, because it grants the com-

Fig. 3.1   Mazes used in experiments.

putational substrate immediate access to almost any part of the maze. Physarum 'knows' *a priori* where the exit from the maze is, because the exit is represented by a source of nutrients. So, the only thing Physarum needs to do is to 'link' the central chamber of the maze with the exit site by a thick protoplasmic tube.

3.2   Single-site start

Will Physarum solve a maze, when it is inoculated only in one place, either in the central chamber or outside the maze? Yes.

In our experiments, we preferred not to cut templates but use ready-made 'party-bag' mazes (Fig. 3.1) you can buy in a supermarket cheaply. We detached the cardboard bottom of the maze, removed the ball bearing and filled the plastic maze with agar gel. To prevent plasmodium from climbing over the maze's walls, we smeared edges of the walls with some unpleasant (for plasmodium) substances such as machine oil, petroleum jelly, etc.; in some cases these repellents worked but in others they did not. An oat flake is placed somewhere in the outer channel of the maze, in the case of the central chamber start, or in the central chamber, if plasmodium enters the maze from outside.

The plasmodium placed in the central chamber propagates and branches along possible passages (Fig. 3.2). It explores several potential routes in

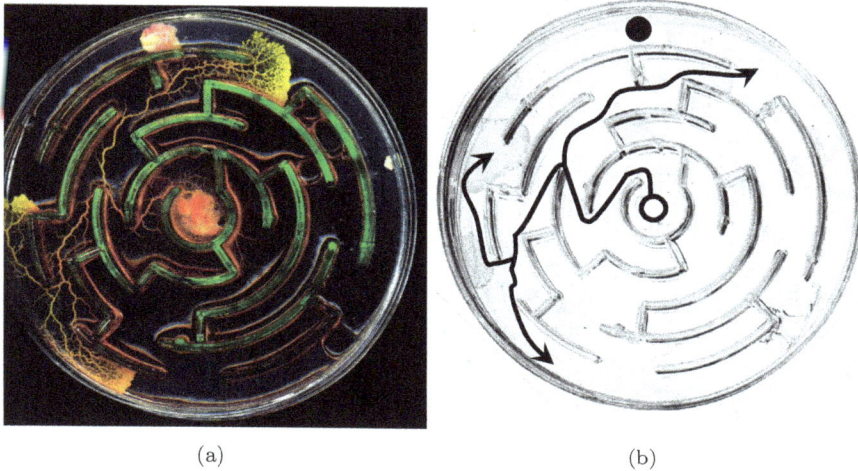

Fig. 3.2 Exploring maze by plasmodium: (a) snapshot of experiment, (b) scheme of the plasmodium, initial site of inoculation is shown by circle, target oat flake by solid disk. Active zones are shown by arrows.

parallel. Active zones of the Physarum machine propagate as typical excitation waves in active nonlinear media, go around the maze wall's corners and leave traces of protoplasmic tubes behind.

Eventually, one active zone of the plasmodium encounters the target source of nutrients (Fig. 3.3). After the target is hit the plasmodium cancels unsuccessful active zones, and retracts or abandons auxiliary protoplasmic tubes. The thickest protoplasmic tube indicates a shortest path from the central chamber to the external source of nutrients. In the example shown in Fig. 3.3, the plasmodium makes 'illegal' moves and climbs over two walls of the maze.

Reversing the locations of the source and destination sites does not confuse the plasmodium. The Physarum machine successfully finds its way into the central chamber, marked with an oat flake, of the maze (Fig. 3.4). When the target source of nutrients is reached, all other active zones are extinguished and their protoplasmic tubes are abandoned.

The Physarum machine never stops (unless deprived of water, oxygen and nutrients). In Fig. 3.5 we see that, 15 h after initialization in the maze, the plasmodium labels the shortest path from the central chamber to the target source of nutrients with a pronounced protoplasmic tube (Fig. 3.5a). The tube marking the shortest path remains stable, and clearly visible, for

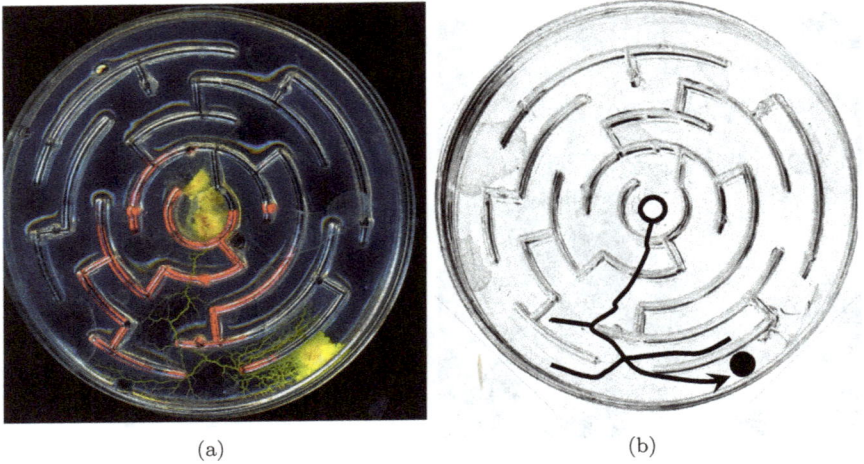

(a)   (b)

Fig. 3.3 Detecting a path from the central chamber to external source of nutrients:
(a) snapshot of experiment, (b) scheme of the plasmodium, initial position of inoculation
is shown by circle, target oat flake by solid disk.

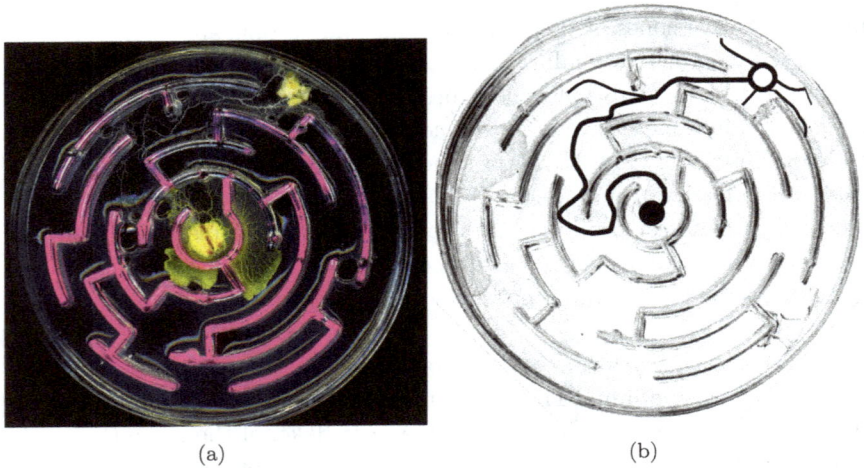

(a)   (b)

Fig. 3.4 Plasmodium constructs a minimal path from outside the maze to the central
chamber: (a) snapshot of experiment, (b) scheme of the plasmodium, initial position of
inoculation is shown by circle, target oat flake by solid disk. Active zones are shown by
arrows and abandoned tubes by lines without arrows.

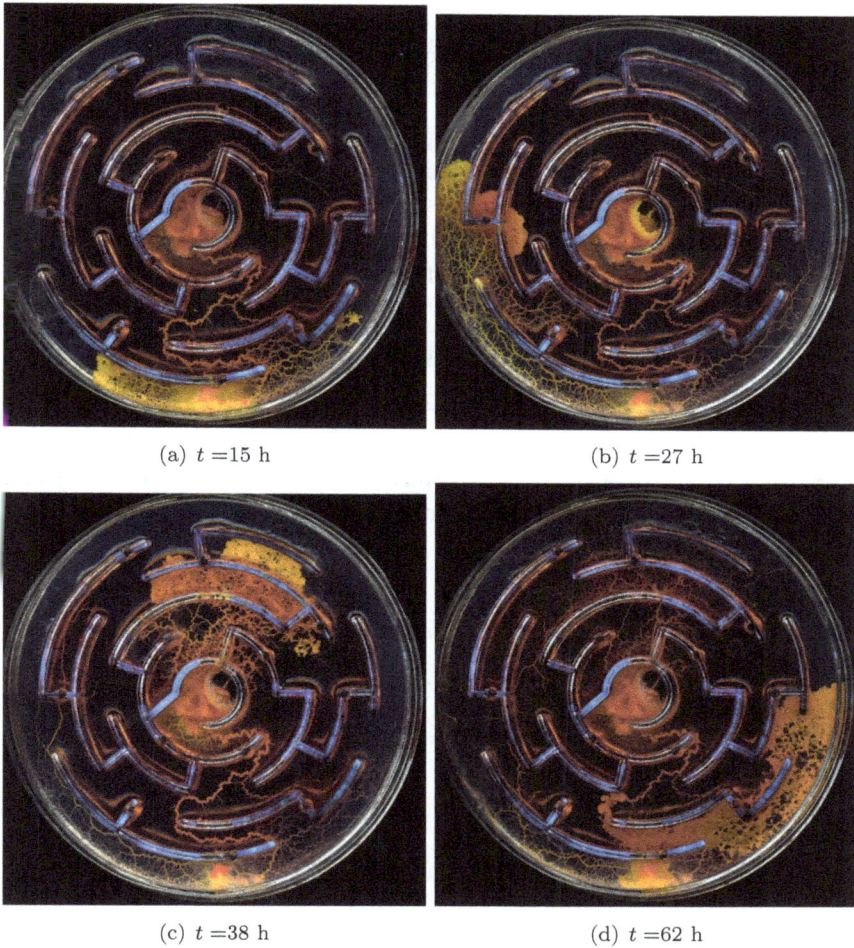

(a) $t = 15$ h

(b) $t = 27$ h

(c) $t = 38$ h

(d) $t = 62$ h

Fig. 3.5 Development of plasmodium in the maze. The plasmodium computes a path from the central chamber to the target oat flake, marks the path with a thick protoplasmic tube (a) and then continues exploration of the maze (b)–(d).

two days (Fig. 3.5d). On reaching the target site the plasmodium does not halt its activity and continues exploration of the maze. In about 24 h it develops active zones in the southern to western parts of the maze (Fig. 3.5b). Then it abandons its south-western activity and propagates north-east (Fig. 3.5c). Finally, almost three days after initial inoculation, the plasmodium checks for nutrients in the south-eastern parts of the maze

(Fig. 3.5d). All this time the main, indicating the shortest route out of the maze, protoplasmic tube remains active and pronounced.

3.3 Summary

We modified the original experiments [Nakagaki et al. (2000, 2001)] and experimentally demonstrated that Physarum machines solve mazes, even when inoculated in one site. Paths represented by Physarum are not minimal. Thus, the main finding of the chapter would be:

Finding 3. *Physarum solves mazes but not necessarily in an optimal way.*

Examples of obstacle-avoiding shortest paths computed by Physarum can be found in almost any chapter of the book, e.g. plasmodium propagates towards oat flakes and avoids salty areas of agar (Chap. 10.10) and plasmodium circumnavigates illuminated domains while reaching oat flakes (Chap. 9.4).

Chapter 4

Plane tessellation

Let $\mathbf{P}$ be a non-empty finite set of planar points. A planar Voronoi diagram of the set $\mathbf{P}$ is a partition of the plane into such regions that, for any element of $\mathbf{P}$, a region corresponding to a unique point p contains all those points of the plane which are closer to p than to any other node of $\mathbf{P}$.

A unique region $vor(p) = \{z \in \mathbf{R}^2 : d(p, z) < d(p, m) \, \forall m \in \mathbf{R}^2, \, m \neq z\}$ assigned to the point p is called a Voronoi cell of the point p. The boundary of the Voronoi cell of the point p is built of segments of bisectors separating pairs of geographically closest points of the given planar set $\mathbf{P}$. A union of all boundaries of the Voronoi cells determines the *planar Voronoi diagram*: $VD(\mathbf{P}) = \cup_{p \in \mathbf{P}} \partial vor(p)$ [Preparata and Shamos (1985)].

Voronoi diagrams are applied in many fields of science and engineering. A few books and conference proceedings are available on the theory and applications of the Voronoi diagram [Okabe et al. (2000); Anton (2009)].

4.1 The ubiquitous diagram

Construction of a Voronoi diagram is a classical problem of unconventional computing devices. This was the first ever problem solved in a reaction–diffusion chemical computer [Tolmachev and Adamatzky (1996)].

The basic concept of constructing Voronoi diagrams with reaction–diffusion systems is based on an intuitive technique for detecting the bisector points separating two given points of the set $\mathbf{P}$. If we drop reagents at the two data points the diffusive waves, or phase waves if the computing substrate is active, travel outwards from the drops. The waves travel the same distance from the sites of origin before they meet one another. The points where the waves meet are the bisector points; see the extensive bibliography in [Adamatzky (2001); Adamatzky et al. (2005)] and mechanisms of bisector formation in chemical media in [De Lacy Costello (2003); De Lacy Costello and Adamatzky (2003); De Lacy Costello et al. (2004,ε, 2009)].

The Voronoi diagram is natural. It represents an optimal space filling. Voronoi diagrams are ubiquitous. They are patterns commonly occurring during point-wise-induced crystallization in spatially extended systems (Fig. 4.1). They are statistical representatives of plant cell tissue [Mebatsion et al. (2006)]. Most growing biological colonies form structures similar to Voronoi diagrams when packed tightly (Fig. 4.2).

The accuracy and time scale of Voronoi diagram approximation vary dramatically between substrates. The diagram is computed in seconds by a crystallization process in a hot ice computer [Adamatzky (2009)] (Fig. 4.1), in hours in chemical precipitating processors [Adamatzky et al. (2005)] and in years in lichen colonies (Figs. 4.2 and 4.3). Chemical processors approximate Voronoi diagrams perfectly (Fig. 4.1), while biological processors are more disorganized (Figs. 4.2 and 4.3).

4.2   Physarum construction of Voronoi diagram

Plasmodium growing on a nutrient substrate from a single site of inoculation expands circularly as a typical diffusive or excitation wave (Fig. 4.4). When two plasmodium waves encounter each other, they stop propagating. To approximate a Voronoi diagram with Physarum, we physically map a configuration of planar data points by inoculating plasmodia on a substrate (Fig. 4.5a). Plasmodium waves propagate circularly from each data point (Fig. 4.5b and c) and stop when they collide with each other (Fig. 4.5d). Thus, the plasmodium waves approximate a Voronoi diagram, whose edges are the substrate's loci not occupied by plasmodia (Figs. 4.5d and 4.6a).

Finding 4. *Plasmodium growing on a nutrient-rich substrate approximates a planar Voronoi diagram.*

The plasmodia do not stop their propagation forever. In 10–15 h they resume their activity, protoplasmic tubes percolate between temporarily 'frozen' wave fronts and the plasmodia originating from different data points merge into a single organism (Fig. 4.6b).

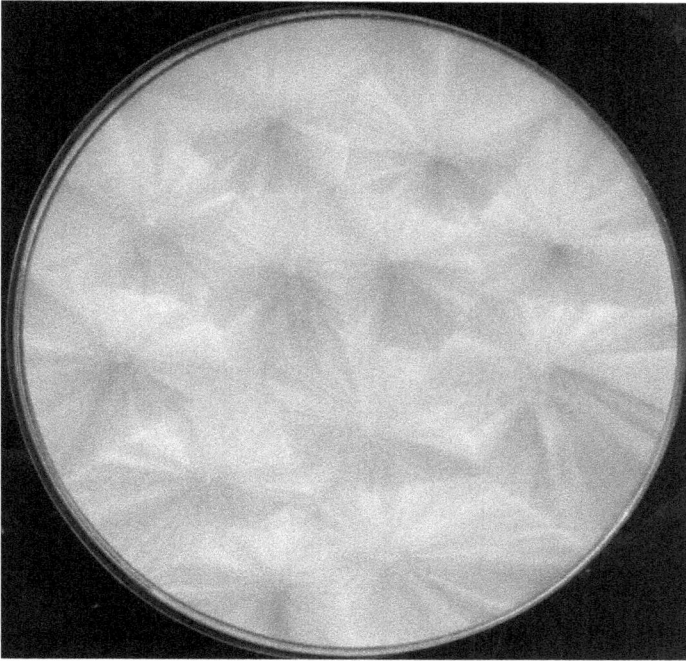

(a)

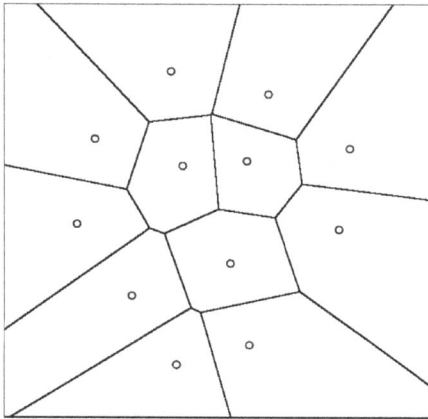

(b)

Fig. 4.1 Voronoi-like tessellation in hot ice computer, constructed during crystallization of supersaturated solution of sodium acetate: (a) photograph of crystallized solution, (b) Voronoi diagram extracted from the image of the crystallized solution perfectly matches the diagram computed by the classical Fortune sweepline algorithm [Fortune (1986)]. See details in [Adamatzky (2009)].

(a)

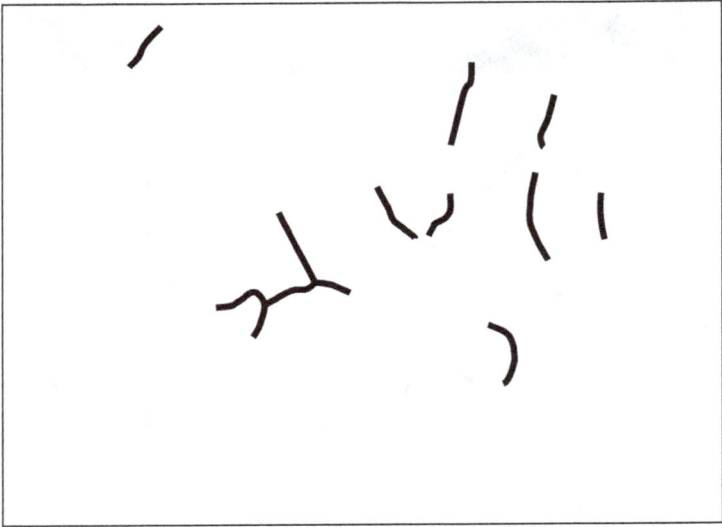

(b)

Fig. 4.2   Voronoi-like tessellation developed in colonies of lichens: (a) photograph of the colonies, (b) bisectors of the tessellation are shown by thick lines.

4.3 Summary

We experimentally demonstrated that plasmodium of *P. polycephalum* approximates a planar Voronoi diagram. The Physarum machine does not store results of computation indefinitely but for a sufficiently long time to detect loci of the substrate representing edges of the Voronoi diagram. Two questions remain unanswered. Why do circular plasmodial waves stop when they collide with each other? How do they sense each other? These may be topics for your home studies.

(a)

(b)

Fig. 4.3   Various stages of Voronoi diagram development in stones colonized by lichens: (a) lichens of two species begin their colonization of the stones, (b) lichens of two species come into direct contact; however, only colonies of the same species form the Voronoi diagram.

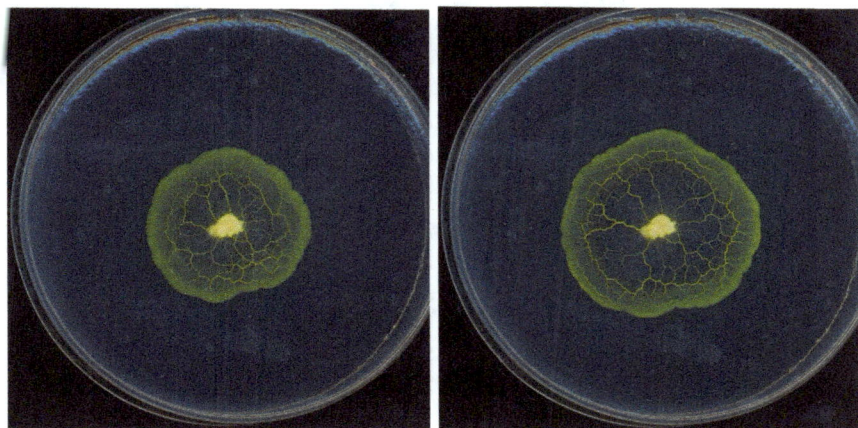

(a)   (b) $t = 9$ h

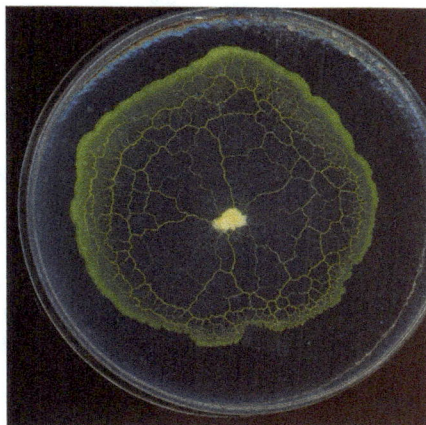

(c) $t = 16$ h

Fig. 4.4   Propagation of Physarum from a single site of inoculation.

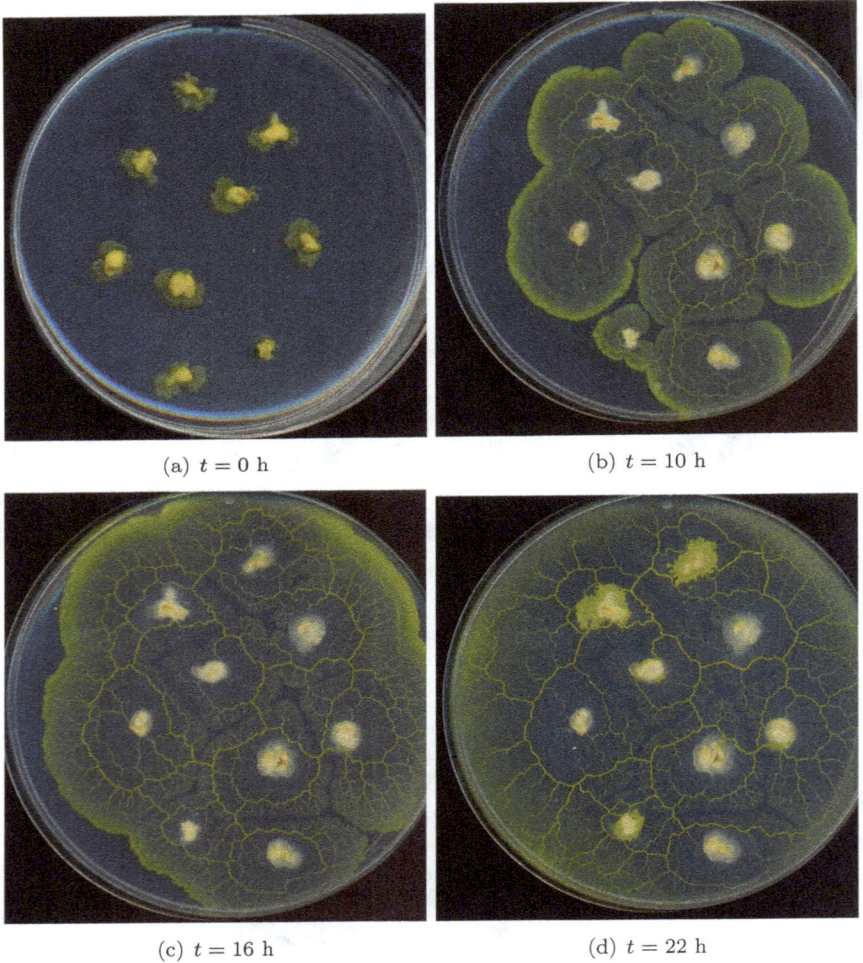

(a) $t = 0$ h

(b) $t = 10$ h

(c) $t = 16$ h

(d) $t = 22$ h

Fig. 4.5 Approximation of Voronoi diagram by Physarum: (a) sites of plasmodium inoculation represent planar data points to be sub-divided by edges of Voronoi diagram, (b) and (c) propagation of growing plasmodia, (d) bisectors of Voronoi diagram are represented by loci of substrate not occupied by plasmodium.

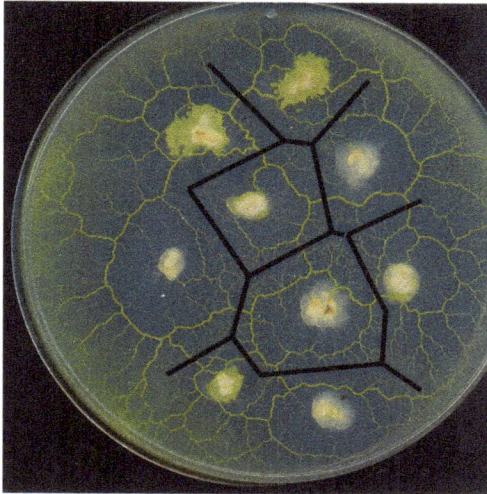

(a) $t = 22$ h

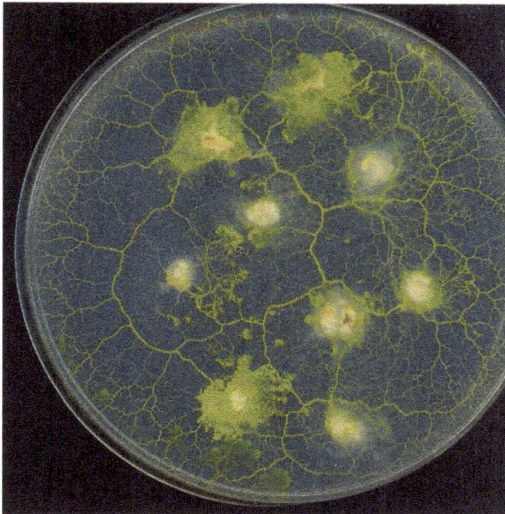

(b) $t = 34$ h

Fig. 4.6 Voronoi diagram approximated by Physarum (a), constructed in 22 h after inoculation. Bisectors of the diagram are represented by loci of substrate not colonized by plasmodium; they are indicated by solid lines. Plasmodia occupy empty loci of the substrate in the next 12 h (b).

Chapter 5

Oregonator model of Physarum growing trees

The main purpose of this chapter is to show that — despite apparent differences between *P. polycephalum* and spatially extended chemical systems — Physarum machines are equivalent to reaction–diffusion chemical computers. We will prove this by simulating Physarum propagation by a two-variable Oregonator model of a Belousov–Zhabotinsky excitable chemical medium. One of the key differences between reaction–diffusion chemical processors and Physarum machines is in how they compute planar graphs [Adamatzky (2007a)]. We take a spanning tree as a benchmark proximity graph.

A spanning tree of a finite planar set is a connected, undirected, acyclic planar graph, whose vertices are points of the planar set; every point of the given planar set is connected to the tree (but no cycles or loops are formed). The tree is a minimal spanning tree whose sum of edge lengths is minimal.

Original algorithms for computing minimum spanning trees are described in [Kruskal (1956); Prim (1957); Dijkstra (1959)]. Hundreds if not thousands of papers were published in the last 50 years, mostly improving the original algorithms or adapting them to multi-processor computing systems. The most common way to construct a spanning tree is to compute a relative neighborhood graph and then to delete some edges of the graph to transform it into a minimal spanning tree [Jaromczyk (1980); Supowit (1988)].

There are some 'unconventional' solutions of the spanning tree problem. Spanning trees can be approximated by random walks, electrical fields, social insects and reaction–diffusion chemical systems [Lyons and Peres (1997); Chong (1993); Adamatzky (2001); Adamatzky and Holland (2002)]. None of these unconventional algorithms however offered an experimental realization.

In 1991, we proposed an algorithm for computing a spanning tree of a finite planar set based on formation of a neurite tree in a development of a single neuron [Adamatzky (1991)]. The algorithm used spreading of neuron processes, and competition between different processes for marked sites of a space was utilized. Morphogenetic and physiological rationales are discussed in [Adamatzky (1991)] and also in [Adamatzky (2001)]. They are based on a range of findings about the behavior of a growing neural tree.

A growth cone as a whole employs a spatial sensing mechanism rather than temporal ones. Filopodia are basic subjects of gradient sensing. Each filopodium of the growth cone uses a temporal gradient. The behavior of each filopodium has a high degree of randomness, which facilitates search of a space. A neurite's branching may be gradient dependent: a growth cone sprouts new filopodia in the direction of maximal concentration of chemo-attractants. Axons may compete for neurotrophins, which are released by targets.

Our idea [Adamatzky (1991)] was to place a neuroblast somewhere on the plane amongst drops of chemical attractants, the positions of which represent points of a given planar set. Then a neurite tree starts to grow. Usually, several growth cones sprout out of the medulloblast body. The growth cones propagate outwards from their sites of origin. Every growth cone probabilistically moves along gradient fields of reagents diffusing from the marked sites of the space. A growth cone can divide itself into several new growth cones. Every growth cone has a wide range of motion and branching modes. Moreover, each growth cone itself decides in which way to elongate and when to branch out and generate a pool of new growth cones, cones of the second order.

At every step of its development, the growth cone can implement one of the procedures: crawl to the next position, generate a limited number of daughter growth cones, detect a point of a given dataset and occupy it (propagate to this point) and resolve conflicts, if any, with other growth cones that try to occupy the same position. We utilize 'energy' — the inverse of lengths of the branches and distances from current positions of growth cones to the neuron soma, or a root of the growing tree, along the branches of the tree — as a main criterion in the competition of several growth cones for the same site of the plane. A cone with higher energy wins. Growth cones can be destroyed not only in competition with other growth cones but also if they do not find a 'target' during some period of time. The algorithms work perfectly on large datasets [Adamatzky (1991, 2007)].

Due to certain circumstances, experimental implementation of the algorithm was not possible at the time of its theoretical investigation [Adamatzky (1991)]. Recent experimental developments in culturing *P. Polycephalum* [Nakagaki et al. (2001); Tsuda et al. (2006); Aono and Gunji (2001, 2004); Tsuda et al. (2004)] convinced us that our algorithm for growing spanning trees can be implemented by living plasmodium.

5.1 What a BZ medium could not do

Proposition 5.1. *Given a finite set **P** of planar points, Physarum can compute a spanning tree of **P** while a Belousov–Zhabotinsky medium could not.*

The proposition assumes that edges of a spanning tree are represented by changes in physical characteristics of the computing substrate. If such an assumption is removed, reaction–diffusion chemical processors would become as potent as Physarum machines.

Assumption 5.1. If wave fragments in a sub-excitable BZ medium did leave permanent traces, then the BZ medium would compute the same spanning trees as Physarum does.

Excitation waves in a sub-excitable BZ medium behave similarly to pseudopodia of *P. polycephalum*, when sources of nutrients, chemo-attractants for the plasmodium, are represented by illumination gradients, 'photo-attractants' for excitation waves, in the BZ medium. The wave fragments are capable of navigating shortest paths towards the source of attraction. Moreover, trajectories of the wave fragments represent edges of a tree spanning the attractant points.

5.2 Physarum and Oregonator

Experiments on growing spanning trees were undertaken in standard Petri dishes, 9 cm in diameter. A substrate was a wet filter paper. We preferred the filter paper and not 2% agar gel in these sets of the experiments, because the paper offers less favorable conditions for the plasmodium growth, and thus less branching of the propagating pseudopodia is observed (see

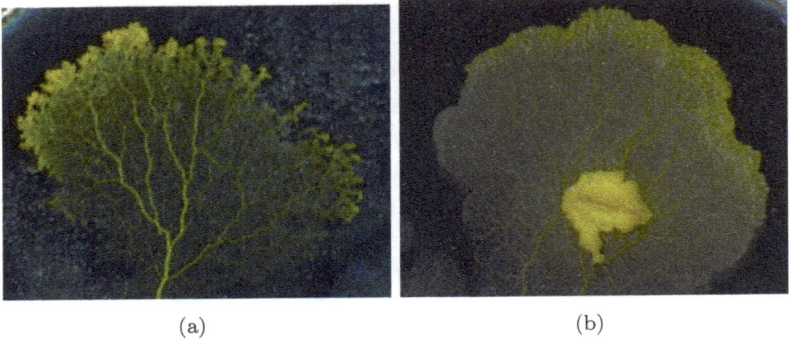

(a)   (b)

Fig. 5.1   Photographs of propagating pseudopodia of a plasmodium of *P. polycephalum*:
(a) pseudopodium propagates following gradients of chemo-attractants and humidity,
(b) the pseudopodium just occupied the oat flake and started to expand further.

details in [Adamatzky et al. (2008); Takamatsu (2007)]). The only dis-
advantage, compared to agar gel, of the filter paper is that it must be
regularly rehydrated; however, this did not make experimentation difficult.
The Petri dishes with plasmodia were kept in darkness and only exposed
to light during observation and recording of images.

Data points, to be connected by protoplasmic graphs, were represented
by oat flakes. Photographs of the protoplasmic networks, developed by the
plasmodium, were made using a FinePix S6500 digital camera.

The profile of the pseudopodium's tip (Fig. 5.1) is isomorphic to shapes
of wave fragments in sub-excitable media. When the pseudopodium propagates, two processes occur simultaneously — propagation of the wave-
shaped tip of the pseudopodium and formation of the trail of protoplasmic
tubes. Examples are shown in Fig. 5.1a and b. We simulate the first process
— tactic traveling of pseudopodia — by excitation wave fragments. The
second process is imitated by an erosion operation applied to the history of
excitation in the medium.

The two-variable Oregonator equations [Field and Noyes (1974); Tyson
and Fife (1980)] adapted to a light-sensitive BZ reaction with applied
gradients of illumination [Beato and Engle (2003)] are as follows:

$$\frac{\partial u}{\partial t} = \frac{1}{\epsilon}\left(u - u^2 - (fv + \phi)\frac{u-q}{u+q}\right) + D_u\nabla^2 u,$$

$$\frac{\partial v}{\partial t} = u - v.$$

The variables u and v represent local concentrations of bromous acid HBrO$_2$ and the oxidized form of the catalyst ruthenium Ru(III). With regard to plasmodium of *P. polycephalum*, the activator, u, is analogous to concentration, or 'thickness', of the plasmodium's cytoplasm at the propagating pseudopodium. The inhibitor, v, combines several factors, where plasmodium is concerned. These factors include rate of nutrient consumption, byproducts of biochemical chains ignited by signals on photo- and chemo-receptors, and concentrations of metabolites released by the plasmodium into its substrate.

The parameter ϵ sets up a ratio of time scales of the variables u and v. In Chap. 6.7.4, we are manually controlling ϵ in our designs of two-input two-output Physarum gates and a one-bit half-adder.

The parameter q is a scaling parameter depending on reaction rates; f is a stoichiometric coefficient.

The parameter ϕ is a light-induced bromide production rate proportional to the intensity of illumination. ϕ is an excitability parameter. A moderate intensity of light will facilitate the excitation process; a higher intensity will produce excessive quantities of bromide, which suppress the reaction. In terms of the plasmodium, ϕ represents the rate of inhibitor proportional to the concentration of nutrients, metabolites, illumination and chemical repellents. See a detailed comparison of BZ and Physarum machines in [Adamatzky et al. (2008); Adamatzky (2009)].

There is no diffusion term for v because we assume that the catalyst is immobilized. To integrate the system, we used the Euler method with a five-node Laplacian operator, time step $\Delta t = 5 \cdot 10^{-3}$ and grid point spacing $\Delta x = 0.25$ (equivalent to 0.6 mm of physical space in terms of plasmodium), with the following parameters: $\phi = \phi_0 - \alpha/2$, $A = 0.0011109$, $\phi_0 = 0.0766$, $\epsilon = 0.03$, $f = 1.4$, $q = 0.022$.

The parameter α imitates the source-dependent gradient of illumination in the BZ medium, which corresponds to a gradient of chemo-attractants emitted by sources of nutrients in the plasmodium's environment. Let **P** be a set of attraction sites, sources of nutrients, and x be a site of a simulated medium; then $\alpha_x = 2 \cdot 10^{-2} - \min_{p \in \mathbf{P}} \{d(x,p) : \gamma(p) = \text{TRUE}\} \cdot b^{-1}$, where $3100 \leq b \leq 4900$ and $d(x,p)$ (for the simulated medium 400×400 sites) is a Euclidean distance between the sites x and p. The parameter $\gamma(p) \in \{\text{TRUE, FALSE}\}$ characterizes the activity of the attraction site (a source of nutrients). At the beginning of simulation $\gamma(p) = \text{TRUE}$ for all sites of **P**. A site p becomes inactive as soon as it is covered by an excitation wave front: if $u_p > 0.1$, then $\gamma(p) = \text{FALSE}$.

Finding 5. 'Extinguishing' of chemo-attractant sources is an important feature of the model. A source of nutrients covered by the plasmodium's pseudopodium ceases releasing chemo-attractants into the surrounding substrate, and thus does not influence the behavior of other pseudopodia.

The medium is perturbed by an initial excitation, when 11×11 sites are assigned $u = 1.0$ each. The perturbation generates a wave, which later splits into one or more wave fragments when affected by a heterogeneous illumination field α. The simulation ends when all sources of nutrients become inactive: for any $p \in \mathbf{P}$, $\gamma(p)$ =FALSE.

Such simulation setup is enough to imitate propagation of protoplasmic pseudopodia as excitation wave fragments and to study navigation of pseudopodia in gradient fields of chemo-attractants. Additional constructs are required to simulate transformation of the pseudopodium's tip to protoplasmic tubes.

During the simulation, values of the variable u are stored in a matrix $\mathbf{L}$. For any site x and time step t, if $u_x > 0.1$ and $L_x = 0$, then $L_x = 1$. At the end of simulation, we repeatedly apply the erosion operation to the matrix $\mathbf{L}$. For any site x, if $L_x = 1$ and $m_1 \leq \sigma_x < m_2$, then $L_x = 0$, where σ_x is a sum of values of x's neighbors in a neighborhood of radius 2: $m_1 = 13$, $m_2 = 15$.

What is the rationale behind the erosion operation leading to simulated formation of protoplasmic tubes? We speculate that the erosion operation represents the stretch-activation effect [Kamiya (1959)]. When repeatedly applied to the trace of the matrix $\mathbf{L}$ of the propagating wave fronts, the operation selects trajectories of those parts of the wave fronts which have maximum curvature. The maximum curvature corresponds to a higher speed of the excitation propagation. This can be expressed as the highest frequency of physico-chemical oscillations of the faster propagating parts of the plasmodium's pseudopodia. The increased frequency evokes a stretch-activation effect on the plasmodium part, which leads to formation of tubular structures [Tero et al. (2007)].

5.3   Building trees with Oregonator

We distribute a few sources of nutrients, sites of $\mathbf{P}$, on the simulated medium. The sources generate gradients of diffusing chemo-attractants. A concentration of 'virtual' chemo-attractant corresponds to an illumination gradient ϕ in the Oregonator model (Fig. 5.3a).

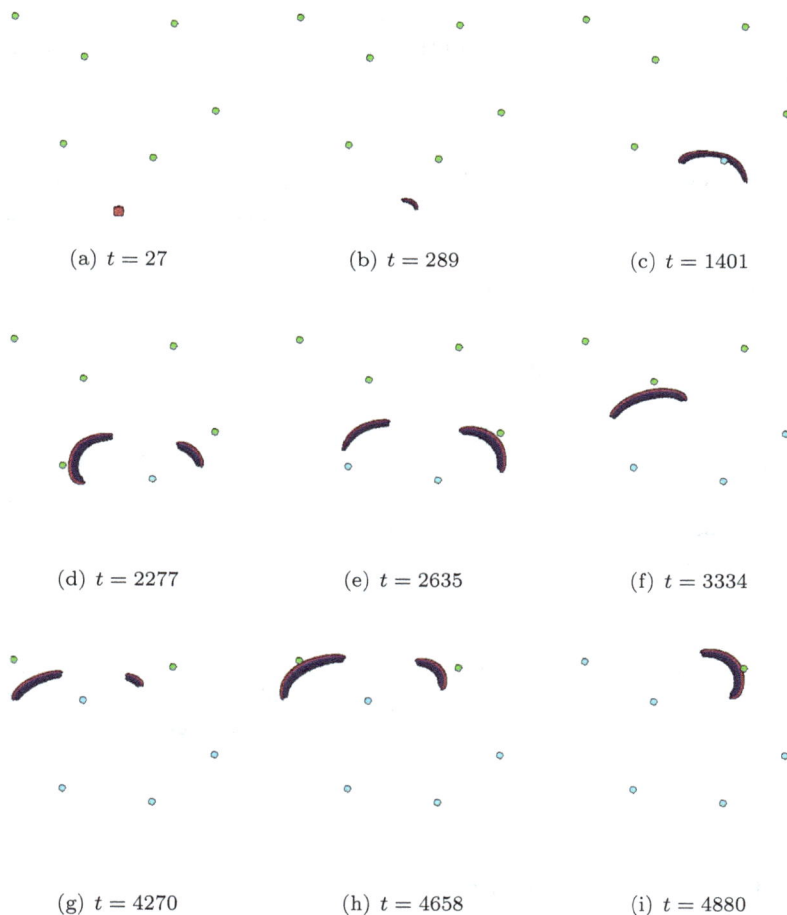

(a) $t = 27$ (b) $t = 289$ (c) $t = 1401$

(d) $t = 2277$ (e) $t = 2635$ (f) $t = 3334$

(g) $t = 4270$ (h) $t = 4658$ (i) $t = 4880$

Fig. 5.2 Snapshots of excitation dynamics taken at various steps t of integration. Sources of nutrients are shown by solid disks. Red and blue components of each pixel's color are defined as follows. If $u > 0.1$, then red value $255 \cdot u$, if $v > 0.1$, then blue value $600 \cdot v$, otherwise the background white color is used.

The medium is perturbed by a small excitation (Fig. 5.2, $t = 27$). An excitation wave fragment is formed and travels towards the closest attractor site (Fig. 5.2, $t = 289$). When a source of nutrients is reached by the wave fragment (Fig. 5.2, $t = 1401$), the source ceases emission of chemo-attractants. The wave fragment splits into two autonomous fragments; one travels to the closest source in the west, the other heads towards the closest north-western source (Fig. 5.2, $t = 2277$). Wave fragments continue to 'extinguish' sources of nutrients until the last attracting site is occupied by an excitation (Fig. 5.2, $t = 4880$). Time lapses of the excitation dynamics are shown in Fig. 5.3b.

After the last source of nutrients is covered by the excitation wave, we apply an erosion operation to all elements of the matrix **L** in parallel. The matrix evolves until it reaches a stable state, where no elements change their values. The stationary form of the matrix **L** (Fig. 5.3cd) approximates a spanning tree of the given planar set **P**.

5.4   Validating simulation by experiments

We arranged oat flakes, on a humid filter paper, in the same configuration as simulated sources of nutrients in Fig. 5.2a scaled up to a Petri dish, 9 cm in diameter. We placed a piece of plasmodium in the southern-most data point, corresponding to the originally excited site in Fig. 5.2a.

Results of one of the experiments are shown in Fig. 5.4. In a few hours after inoculation (Fig. 5.4a), the plasmodium propagates towards the north-eastern oat flake, closest to the site of initial inoculation. It then continues its foraging behavior until all sources of nutrients are occupied and linked together with protoplasmic tubes (Fig. 5.4b and c). The final graph of protoplasmic tubes, enhanced by standard procedures, is shown in Fig. 5.4d.

5.5   Summary

Ideas discussed in the chapter emerged from our early computational and laboratory experiments on developing unconventional computing circuits using operations with wave fragments in a BZ medium [Adamatzky (2004); De Lacy Costello and Adamatzky (2005)] and the fact that wave fragments in a sub-excitable BZ medium behave similarly to pseudopodia of the plasmodium cultivated on a nutrient-poor substrate [Adamatzky et al. (2008)].

In numerical experiments, we demonstrated that propagation of wave fragments in a sub-excitable light-sensitive BZ medium imitates foraging behavior of pseudopodia of plasmodium of *P. polycephalum* when a chemo-attractant field generated by sources of nutrients is represented by gradients of illumination in the BZ medium. The traveling wave fragments compute the spanning tree of the attractor sites, or domains of lower illumination.

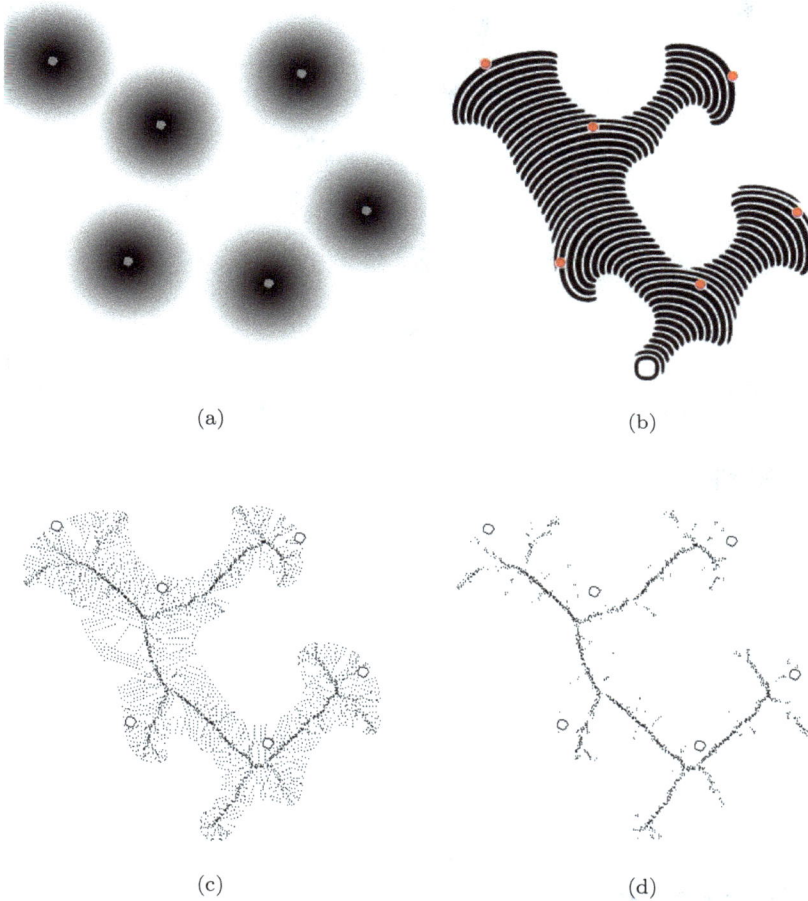

Fig. 5.3 History of excitation and the spanning tree approximated. (a) Visualization of attractant concentration, or illumination, in the simulated medium. Gray level at each pixel x is calculated as $255 \cdot (1 - \alpha_x \mu^{-1})$, where $\mu = \max\{\alpha_x | x \in \mathbf{L}^2, p \in \mathbf{P}\}$. Sources of nutrients are shown by gray disks centered in their gradient fields. (b) Overlaid images of propagating wave fronts; overlay of images is taken every 100 steps of integration. (c) Stationary form of the matrix $\mathbf{L}$ after repeated application of the erosion operation. (d) The matrix $\mathbf{L}$ after additional removal of pixels, which have fewer than three non-white neighbors. Sources of nutrients are shown by circles.

Results of our experiments complement previous findings on plasmodium behavior that the plasmodium moves towards stimulants that increase its frequency of protoplasm streaming [Durham and Ridgway

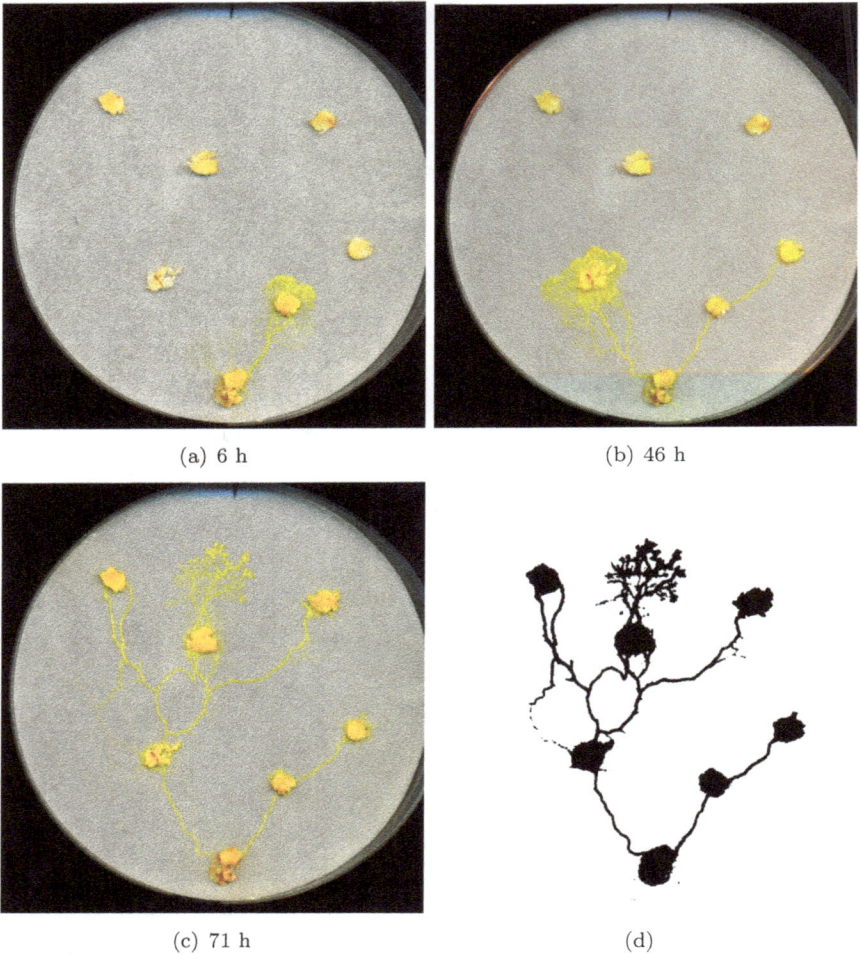

(a) 6 h

(b) 46 h

(c) 71 h

(d)

Fig. 5.4 Photographs of the living plasmodium building a spanning tree on the planar configuration of oat flakes: (a)–(c) snapshots of the Petri dish with plasmodium are taken at 6, 46 and 71 h after inoculation of the plasmodium; (d) network of protoplasmic tubes from (c) is enhanced by standard procedures.

(1976)], determined by chemo-induced oscillations of calcium-ion concentrations [Ridgway and Durhma (1976); Rapp (1976)]. The results also shed light onto spatial development of the plasmodium. Previously published models dealt with simulation of the plasmodium as coupled mechanical [Romanovskii and Teplov (1995); Takahashi et al. (1997); Takamatsu et al. (2000); Tero et al. (2005); Takamatsu (2006)] and chemical oscillators [Ueda (1993)], where signal transmission in response to stimulation is induced by a shift of oscillation [Achenbach and Wolfarth-Bottermann (1980); Nakagaki et al (1999)]. These models did not give an adequate answer to how exactly propagation of the plasmodium's pseudopodia occurs. Using the two-variable Oregonator model of the BZ system, we demonstrated that the behavior of the pseudopodia can be satisfactorily imitated by propagating wave fragments, and that trajectories of the pseudopodium propagation are determined mostly by a configuration of chemo-attractant gradient fields.

In contrast to the simulated BZ system, the plasmodium rarely stops foraging even when the spanning tree of sources of nutrients is constructed. The plasmodium continues developing its protoplasmic network connecting existing edges by its tube of protoplasm [Adamatzky (2008)]. In terms of the BZ system, the continued development can be imitated by perturbing all sites of $\mathbf{P}$, and allowing for asynchronous excitation of the loci. This may be a subject of further investigation. We are particularly curious to know whether the BZ system can simulate transition of planar graphs from the spanning tree to a relative neighborhood graph and to a Gabriel graph as well as a Physarum machine does [Adamatzky (2008)] or not.

A biological analog of the erosion of the matrix $\mathbf{L}$ is yet another unclear point. We envisage that protoplasmic tubes are formed at the places of highest oscillatory activity of the pseudopodium's loci. These regions may correspond to higher values of u in the propagating excitation wave fronts. Further laboratory experiments are required to substantiate this hypothesis.

Chapter 6

Does the plasmodium follow Toussaint hierarchy?

Networks of protoplasmic tubes developed by plasmodium of *P. poly-cephalum* are amazing in their beauty, complexity and transport optimality. When two protoplasmic tubes meet they usually merge, forming a ⊐-junction; therefore, the majority of networks generated by the plasmodium are planar graphs. Growing pseudopodia rely mainly on chemical, light and moisture gradients when choosing the direction of their growth. Therefore, we can assume that *protoplasmic graphs constructed by plasmodium belong to the family of proximity graphs* [Jaromczyk and Toussaint (1992)].

It is well known from the works [Nakagaki et al. (2000, 2001)] that the plasmodium's protoplasmic network develops in such a manner as to optimize harvesting of distributed sources of nutrients and to make more efficient flow and transport of intracellular components. However, the questions remain 'up to what degree do the protoplasmic networks approximate optimal planar graphs spanning the sources of nutrients?', 'what exact types of graphs are approximated?' and 'how does the topology of the protoplasmic graphs change during the plasmodium's development?'.

6.1 Proximity graphs

To construct a proximity graph of a set of points, one selects a measure of neighborhood and closeness $\mathcal{C}$ and then connects those points which are close neighbors in $\mathcal{C}$ [Toussaint (1980, 1989); Jaromczyk and Toussaint (1992)] with edges of the graph.

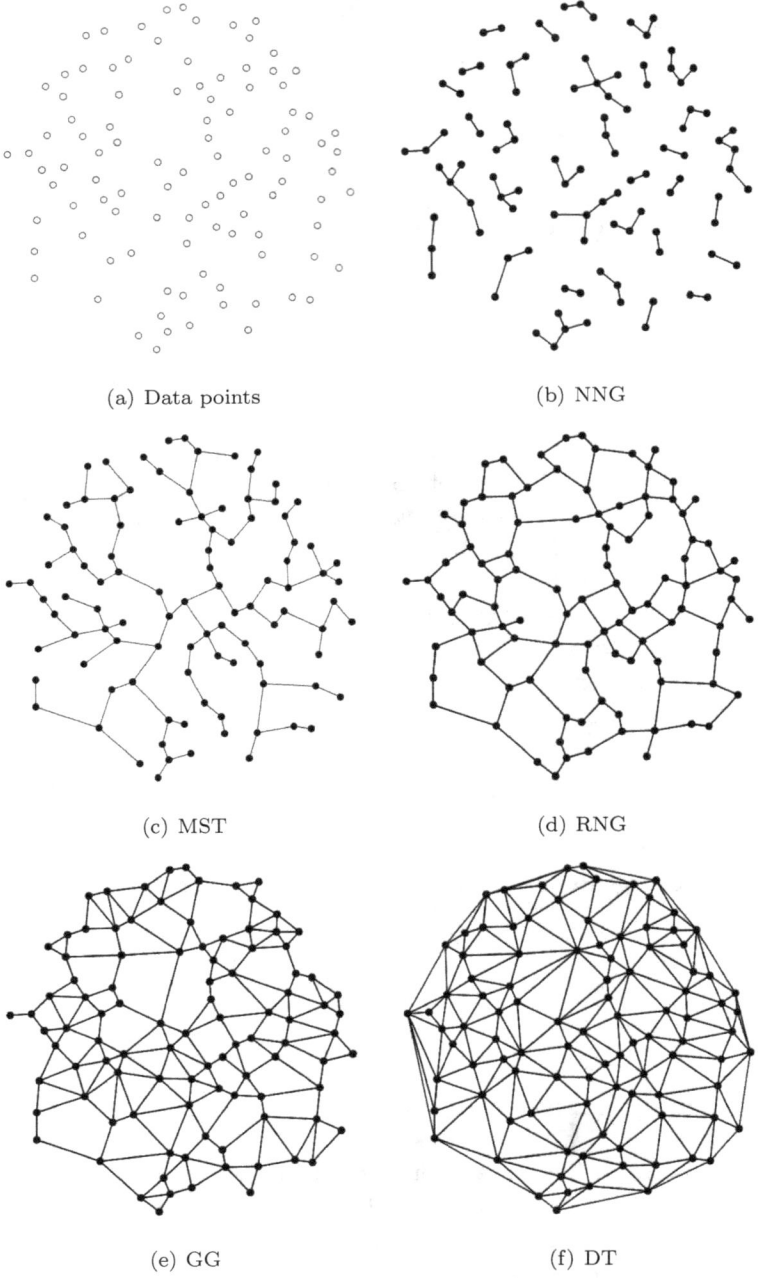

(a) Data points

(b) NNG

(c) MST

(d) RNG

(e) GG

(f) DT

Fig. 6.1   Examples of proximity graphs.

6.1.1   Nearest-neighborhood graph

A nearest-neighborhood graph (NNG) is the simplest and possibly most natural of the proximity graphs. A point in the graph is connected by an edge to its nearest neighbor [Preparata and Shamos (1985); Paterson and Yao (1992)] (Fig. 6.1b). Given a planar set $\mathbf{V}$, we can define the graph as follows: $\mathrm{NNG}(\mathbf{V}) = \langle \mathbf{V}, \mathbf{E} \rangle$, where, for $a, b \in \mathbf{V}$, we have $(ab) \in \mathbf{E}$ if and only if $|ab| = \min_{c \in \mathbf{V} - \{\mathbf{a}\}} |ac|$. In the general case, a NNG is a disconnected directed graph. All other graphs discussed here are undirected.

6.1.2   Minimal spanning tree

The Euclidean minimal spanning tree (MST) [Nesetril et al. (2001)] is a connected acyclic graph which has minimal possible sum of edges' lengths (Fig. 6.1c).

6.1.3   Relative neighborhood graph

In a relative neighbourhood graph (RNG) [Toussaint (1980)], any two points (a, b) are connected by an edge if the intersection of open disks of radius $|ab|$ centered at a and b is empty (Fig. 6.1d): $(ab) \in \mathbf{E}$ if and only if $|ab| \leq \max_{c \in \mathbf{V} - \{a,b\}} \{|ac|, |bc|\}$.

6.1.4   Gabriel graph

In a Gabriel graph (GG) [Gabriel and Sokal (1969); Matual and Sokal (1984)], points a and b are connected by an edge if the closed disk having the segment (ab) as its diameter is empty (Fig. 6.1e): $(ab) \in \mathbf{E}$ if and only if $|ab| \geq \min_{c \in \mathbf{V} - \{a,b\}} \{|\frac{a+b}{2}c|\}$.

6.1.5   Delaunay triangulation

The Delaunay triangulation (DT) [Delaunay (1934)] is a graph sub-dividing the space into triangles with vertices in $\mathbf{V}$ and edges in $\mathbf{E}$, where the circumcircle of any triangle contains no points of $\mathbf{V}$ other than its vertices (Fig. 6.1f).

6.1.6   *Toussaint hierarchy*

In 1980, Toussaint demonstrated that MST⊆RNG⊆DT [Toussaint (1980)]. This hierarchy of the graphs was later enriched with GG [Matual and Sokal (1984)], and we can add NNG at the lower level of the enclosure hierarchy by default:

$$ \text{NNG} \subseteq \text{MST} \subseteq \text{RNG} \subseteq \text{GG} \subseteq \text{DT}. $$

Amongst other features, the Toussaint hierarchy represents dynamics of partial closure of a graph; starting from a NNG, any next graph in the hierarchy is produced from the previous graph by adding some edges between non-adjacent nodes; moreover, after the transition MST→RNG, decondensation starts because more cycles emerge during the partial closure.

6.2   Plasmodium network and Toussaint hierarchy

Given a planar set **V** represented by sources of nutrients and inoculated with the plasmodium of *P. polycephalum*, with time the plasmodium connects elements of **V** with its network of protoplasmic tubes. Does the dynamics of the topology of the protoplasmic network follow the inclusion hierarchy NNG→MST→RNG→GG→DT?

We hypothesize that the plasmodium, at least when developing from a single point, does follow the Toussaint hierarchy, when spanning other points. When placed on a substrate, the plasmodium is faced with chemical diffusive gradients, generated by bacteria growing on oat flakes or oat flakes themselves. See an example of such a gradient field in Fig. 6.2a. When each next oat flake is covered by the plasmodium, it ceases to generate more chemo-attractants and stops contributing to the gradient field.

The plasmodium is 'amplified' (similar to amplification of signals in electronic circuits) at the sources of nutrients. This is the reason why the whole length of a plasmodium branch, e.g. from the root to the current leaf, may not be that important for the plasmodium. A pseudopodium only makes a choice based on its length from the node it just left.

Immediately on propagating away from its current source of nutrients, a pseudopodium takes a shape close to semicircular (Fig. 6.2b). A wave fragment like shaped growing tip of the pseudopodium interacts with, in an ideal situation, circular shapes of diffusion gradients (Fig. 6.2c). This

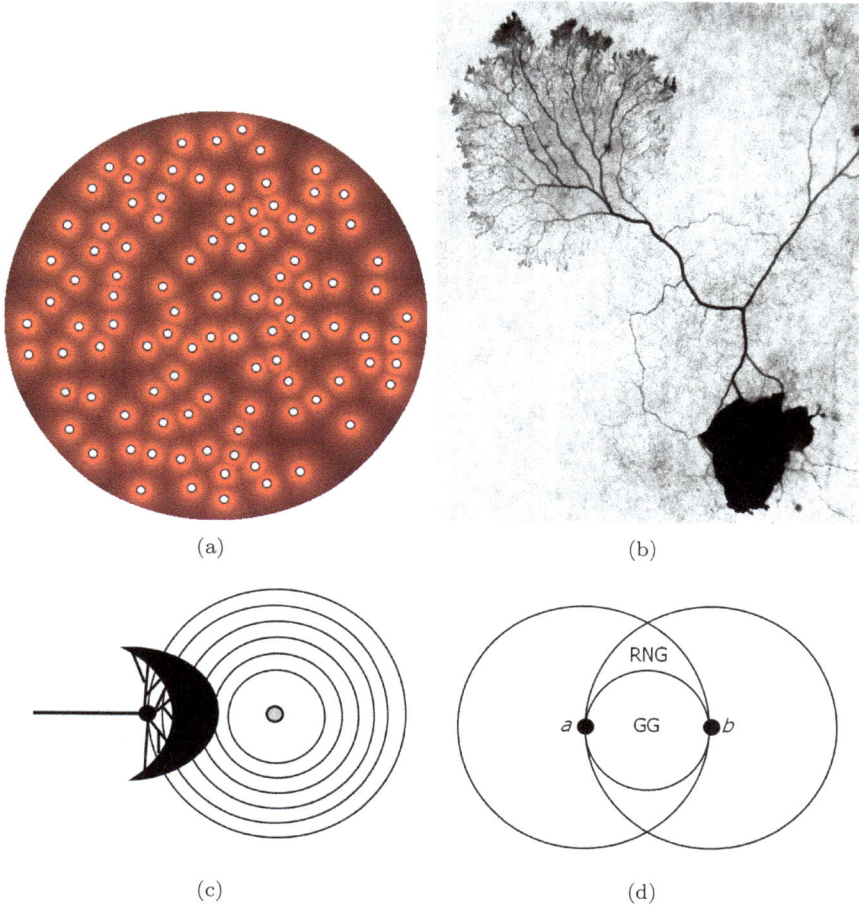

Fig. 6.2   Interaction of pseudopodia with environment: (a) vsualization of possible diffusion gradient guiding the plasmodium. (b) Amplification of pseudopodia at the source of
nutrients. The pseudopodium traveled from the original site, shown by solid black shape
in the bottom right corner of the image, to a drop of agar gel with crushed oat flakes.
The drop of agar with nutrients is faintly visible as a gray disk covered by pseudopodia.
(c) Relation of a pseudopodium's zone of sensitivity and gradients of chemo-attractants
diffusing from the source of nutrients. (d) Relation of lune (RNG) and ball (GG) neigh-
borhoods.

type of interaction reminds us of a lune-based vicinity employed in a RNG
(Fig. 6.2d). So, our assumption would be that at first the plasmodium
constructs a spanning tree (adding edges which satisfy RNG conditions

but preventing formation of cycles in the graph). Then, when each data point is covered by plasmodium, the plasmodium nodes establish additional connections to form a RNG. At later stages more protoplasmic tubes are added, to increase reliability and fault tolerance. Thus, a GG and then a DT are formed.

It would also be interesting to check whether or not dynamics of a protoplasmic tree built by plasmodium, when the plasmodium starts its growth in a single site, correspond to space–time dynamics of growing, also from a single point, a spanning tree based on a neuroblast growth algorithm [Adamatzky (1991)] and where next neighbors are selected on lune-neighbourhood conditions (as in a RNG).

See an example of a spanning tree of a large number of nodes, growing from a single point, in Fig. 6.3. Does the plasmodium grow similar trees?

6.3   Preparing for graph growing

Experiments on construction of planar graphs by the plasmodium were undertaken in standard Petri dishes, 9 cm in diameter. A substrate was a wet filter paper or, in a few cases, a 2% non-nutrient agar plate.. The Petri dishes with plasmodia were kept in darkness and only exposed to light during observation and recording of images.

Data points, to be connected by protoplasmic graphs, were represented by either oat flakes or drops of agar gel with crushed oat flakes. The former are more visible in the photographs, the latter offer more accurate shapes. Photographs of the protoplasmic networks, constructed by the plasmodium, were made using a FinePix S6500 digital camera. The images were overlaid with known types of proximity graphs calculated by standard algorithms. The software was written in Processing[1].

Two kinds of experiments were undertaken:

- growing plasmodium graphs from a single point: one piece of plasmodium is placed on a substrate populated with oat flakes, and
- simultaneous growing from all data points: each point of a given planar set was represented by a piece of plasmodium.

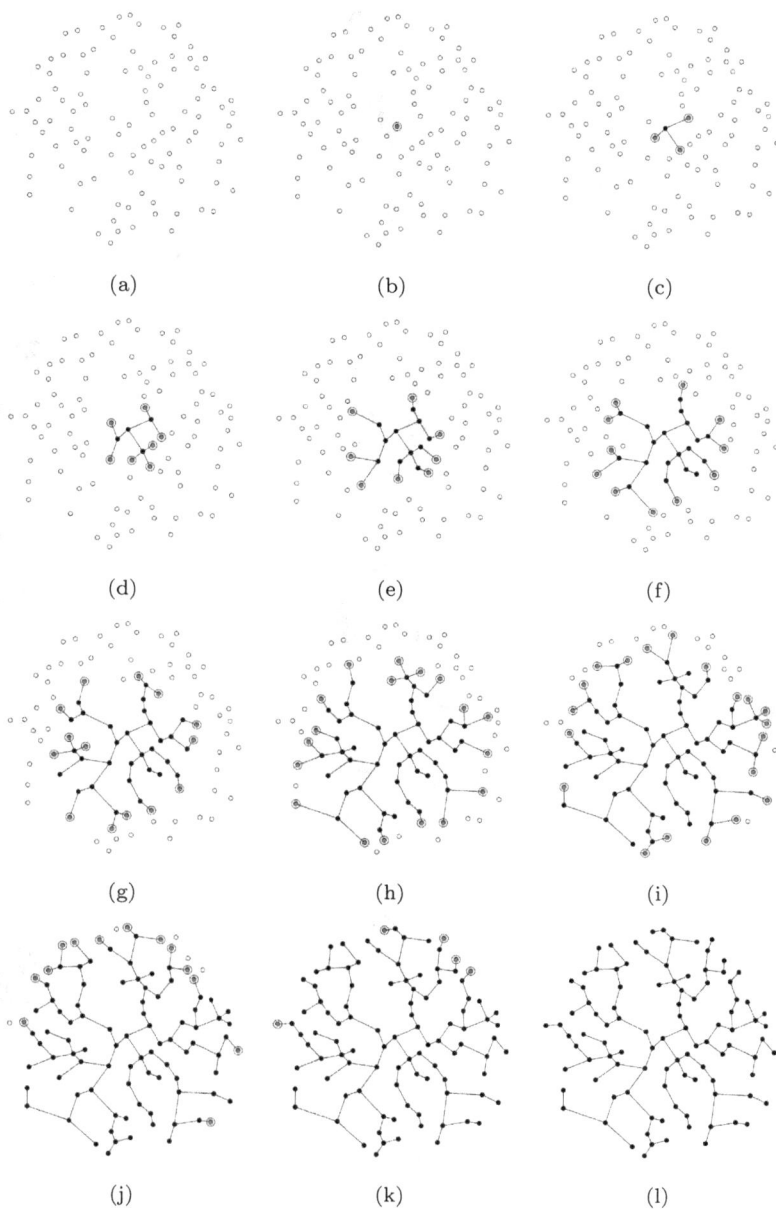

Fig. 6.3 Spanning tree grown from a single point. 100 planar points are randomly distributed in a unit disk. Data points not incorporated in the graph and not activated are shown by circles, points spanned by the tree are solid disks and 'active' points, where pseudopodia reside, are shown as small solid disks inside circles.

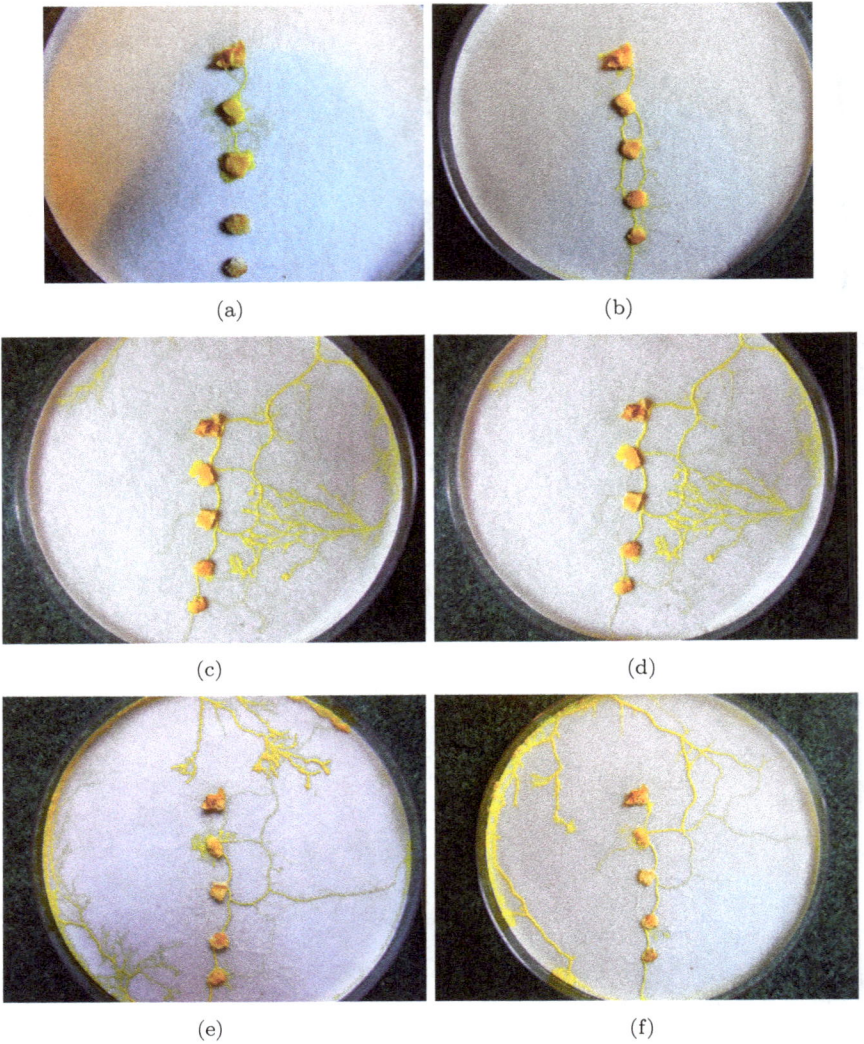

Fig. 6.4   Spanning a chain by protoplasmic network.

6.4   Growing graph from a single point

The simplest ever experiment is to arrange data points along a straight line
and then place the plasmodium at one of the terminal sites (Fig. 6.4). Then

---

[1]www.processing.org

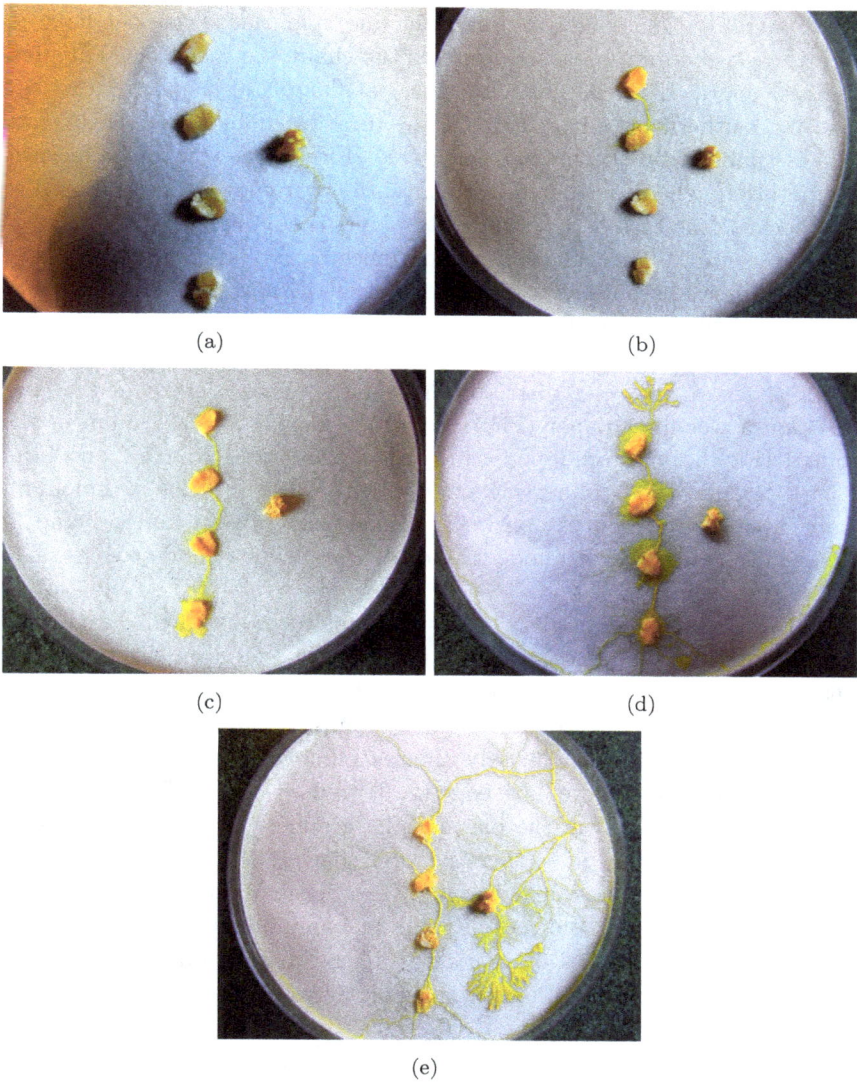

Fig. 6.5 Spanning a chain of planar sites by a protoplasmic network starting from outside the chain. Initially, the plasmodium is on the eastern-most flake.

the plasmodium gradually constructs a NNG by connecting points one by one in a sequence. When all points are connected by the plasmodium's protoplasmic tubes, the network represents a MST.

We arranged oat flakes in a straight line, and placed one oat flake — with the plasmodium on top — outside the chain, see Fig. 6.5a. The plasmodium migrated to the northern-most oat flake, and connected this site with its southern neighbor (Fig. 6.5b). In a few hops the chain of data sites is spanned by the protoplasmic tree (Fig. 6.5c). At this stage, the computation, when considered in a conventional framework, is supposed to stop because the task of spanning data points is completed. The plasmodium never stops. The plasmodium continues to explore the space. Thus, Fig. 6.5d shows pseudopodia spreading out of the northern-most site until the original site (where the plasmodium was inoculated initially) is added to the protoplasmic network (Fig. 6.5e).

Consider an irregular distribution of data points, represented by oat flakes on a wet filter paper (Fig. 6.6a). We place the plasmodium in the point 1 (Fig. 6.6b). Pseudopodia propagate from site 1 to site 2 and form a protoplasmic tube connecting node 1 with node 2. The plasmodium propagates further and establishes s connection between nodes 2 and 3, then connects node 3 with node 4, and then branches at node 4 to develop edges (4, 6) and (4, 5).

In this particular example, the plasmodium does not span all data nodes, so it just partially constructs a spanning tree of the given set. Does the plasmodium at least construct a minimum tree of the nodes { 1, ..., 6 }?

In Fig. 6.6b, we can see a MST of the complete dataset constructed by conventional algorithms. The only difference between the sub-trees is that in the plasmodium tree we have edge (2, 3) while in the MST we have edge (1, 3). The edge (2, 3) in the plasmodium graph is 1.4 times longer — if measured as a straight line — than the MST edge (1, 3). However, a straight line means nothing to the plasmodium. The plasmodium tries to optimize the routes in terms of comfortable humidity and substrate structure, not only proximity to sources of nutrients. If the plasmodium grew a MST properly, then dynamics of its protoplasmic network would be as shown in Fig. 6.6c–g.

Consider an almost regular structure — a grid-like arrangement of oat flakes, with plasmodium placed just below the south-western corner of the grid (Fig. 6.7; the initial position of the plasmodium is marked by 1 in Fig. 6.8. Dynamics of protoplasmic network development are shown in Fig. 6.7. As can be seen in Fig. 6.7d, the plasmodium continues branching additional protoplasmic tubes even after spanning all data points (oat flakes) by a proximity graph.

The basic proximity graphs, computed by standard techniques, are over-

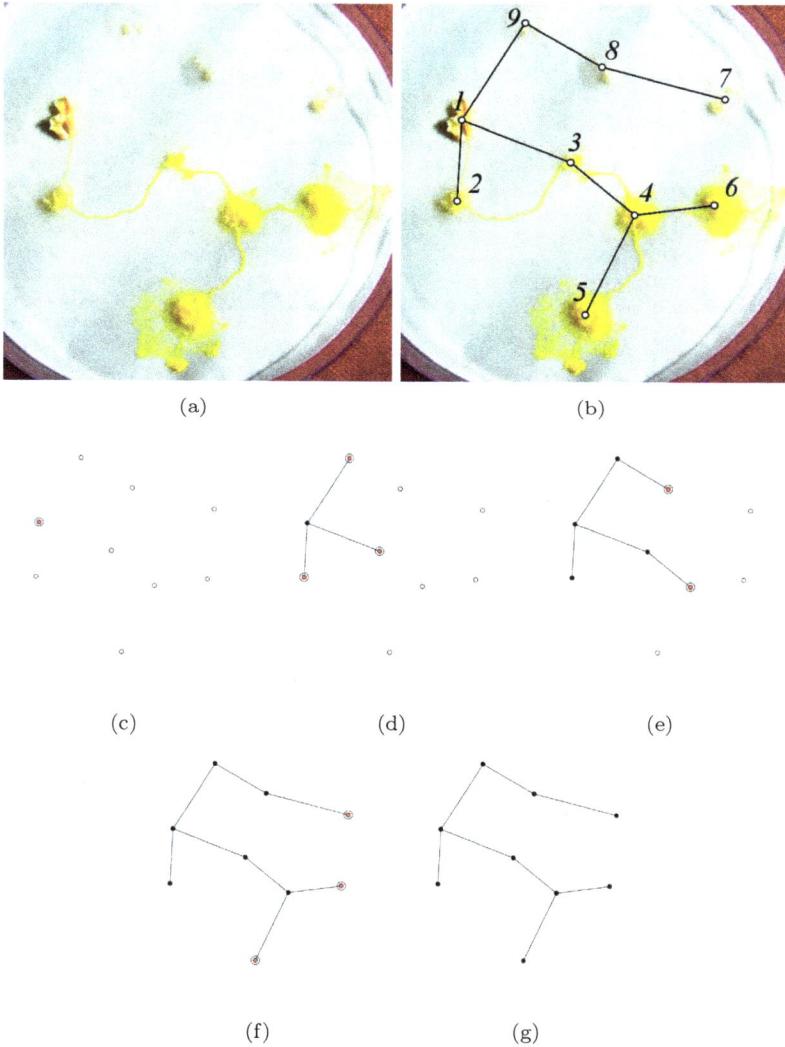

Fig. 6.6 Growing tree with plasmodium: (a) positions of oat flakes representing data points, (b) MST overlaid with snapshot of partially constructed plasmodium tree, (c)–(g) dynamics of growing 'would-be' MST by the plasmodium.

laid with the plasmodium network in Fig. 6.8. The plasmodium approximates a RNG satisfactorily (Fig. 6.8b); just a few edges mismatch. For example, nodes 6 and 10 are connected by a protoplasmic tube but there

Fig. 6.7   Development of protoplasmic network by the plasmodium. Snapshots (a)–(d) are recorded at 10 h intervals.

is no edge (6, 10) in the RNG. The edge (6, 10) appears in the GG; however, the GG looks slightly redundant when compared with the plasmodium network because the edge (2, 6) present in the GG lacks a corresponding protoplasmic tube in the plasmodium network. The DT contains too many 'extra' edges, not present in the plasmodium network.

The observations tell us that in this particular example the *plasmodium approximates a proximity graph lying somewhere between a RNG and a GG*. The graph constructed by the plasmodium does not match β-skeletons [Kirkpatrick and Radke (1985)] either for any values of β.

In the present example, the plasmodium does not approximate a spanning tree completely, but just at first steps of its development (Fig. 6.7a and b). Dynamics of edge formation in the plasmodium network (Fig. 6.7a

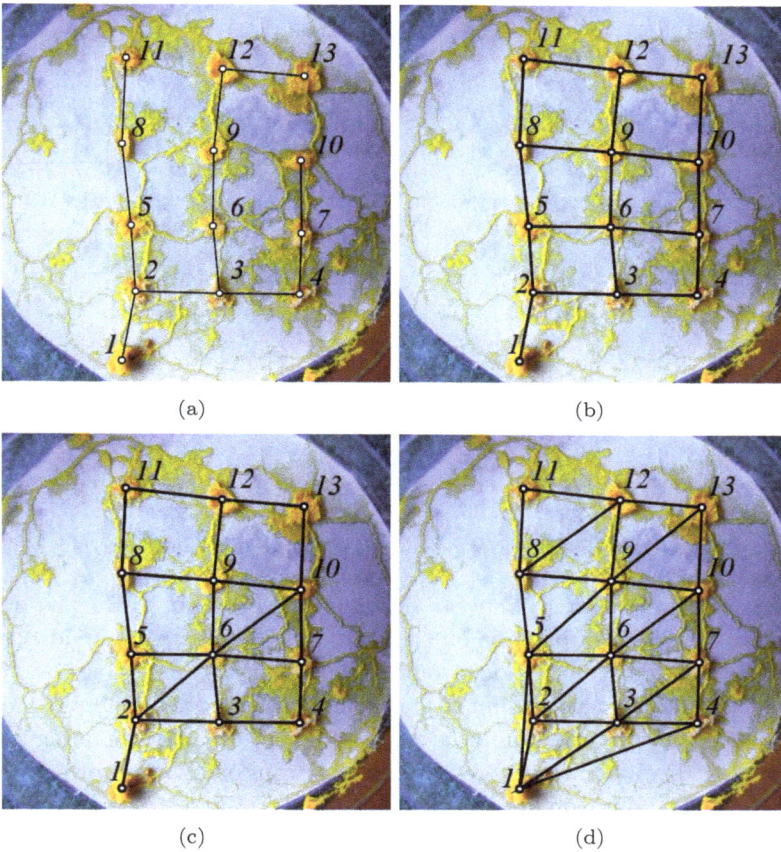

Fig. 6.8 Proximity graphs overlaid with the plasmodium network: (a) MST, (b) RNG, (c) GG, (d) DT.

and b) and the growing MST (Fig. 6.9) are as follows:

Hop time	Edges in plasmodium	Edges in MST
$t = 1$	$(1, 2)$	$(1, 2)$
$t = 2$	$(2, 5), (2, 3)$	$(2, 5), (2, 3)$
$t = 3$	$(5, 8), (3, 4)$	$(5, 8), (3, 6), (3, 4)$
$t = 4$	$(8, 11), (8, 9)$	$(8, 11), (6, 9), (4, 7)$

The plasmodium well approximates a MST at early stages of its development. Later, the plasmodium network deviates from minimality of connections. This may be explained in that, when building a next edge between two sources of nutrients, the plasmodium takes into account not only

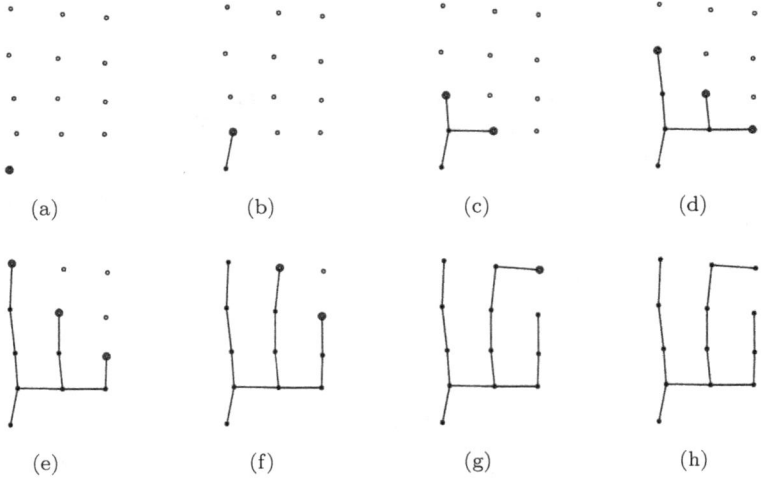

Fig. 6.9   Growing MST on the same dataset as the plasmodium used in Fig. 6.7.

'as the crow flies' distance between neighboring nodes but also many other conditions: amount of bacteria on the oat flakes, gradients of humidity and quality of substrate on the loci around neighboring oat flakes.

Is it typical for plasmodium to construct an incomplete spanning trees? Look at another plasmodium tree compared with the corresponding MST in Fig. 6.10. Data points were represented by drops of 2% agar with crushed oat flakes. The data points are arranged in a grid-like structure of 4×4 points, and the oat flake with the plasmodium is placed in the southern-most site of the Petri dish (Fig. 6.10a). The MST constructed by a standard algorithm (growth from the same site as the plasmodium grew) is overlaid with a photograph of the plasmodium network in Fig. 6.10e. From the snapshots of plasmodium development and growth dynamics of the MST, we can extract the following temporal sequence of edge formation:

Hop time	Edges in plasmodium	Edges in MST
$t = 1$	$(1,3), (1,4)$	$(1,3), (1,4)$
$t = 2$	$(2,3), (4,5), (4,8)$	$(2,3), (4,5), (4,8), (3,7)$
$t = 3$	$(8,12)$	$(8,12), (7,6), (8,9), (7,11)$
$t = 4$	$(12,16), (12,11)$	$(12,16), (11,15), (12,13)$
$t = 5$	$(16,17)$	$(16,17), (15,13)$

We see that the plasmodium-built graph would be a sub-graph of the

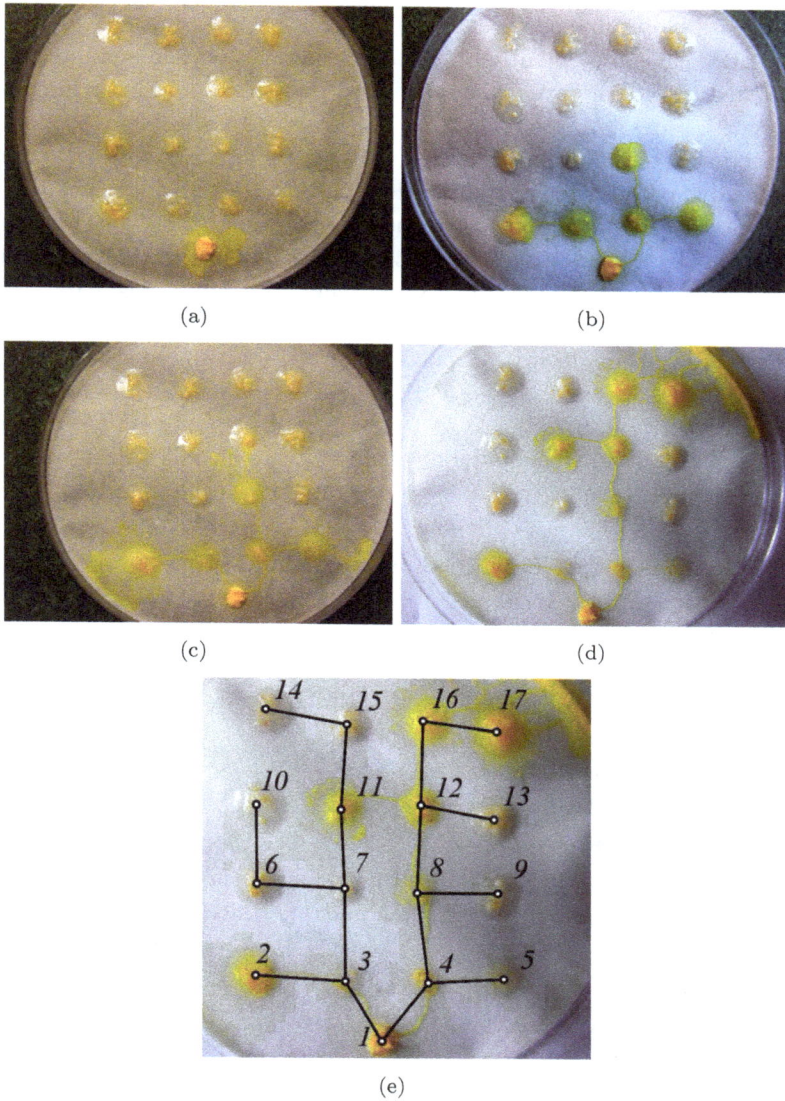

Fig. 6.10 Development of partial spanning tree by the plasmodium: (a)–(d) snapshots of the protoplasmic network taken at 10 h intervals, (e) MST overlaid with the plasmodium network.

MST except for the protoplasmic edge (11, 12). For the rest, the plasmodium well approximates a partial MST. It is unclear why the plasmodium does not develop some edges existing in the MST, e.g. edges (7, 6), (8, 9) and (7, 12). This may be attributed to the fact that at some points only a single pseudopodium develops at each next data point and branching does not occur.

In some cases the plasmodium can construct a tree partially, then abandon already spanned nodes and develop another tree spanning not yet occupied nodes. We exemplify this phenomenon in Fig. 6.11. The tree is constructed by the plasmodium in two phases. During the first phase, nodes 1, 2, 4, 6, 10, 11, 12 and 13 are occupied by the plasmodium and connected by its protoplasmic network (Fig. 6.11a–c); see labeling of nodes in Fig. 6.11e. The nodes 3, 7, 8, 9, 17, 15 and 16 are spanned by the plasmodium during the second phase. Some nodes spanned in the first phase are abandoned (Fig. 6.11d).

This plasmodium tree almost nowhere matches a MST neither in edges nor in a temporal sequence of its construction (Fig. 6.11e and f):

Hop time	Edges in plasmodium	Edges in MST
$t = 1$	$(1,5), (1,4)$	$(1,2), (1,3), (1,4), (1,5)$
$t = 2$	$(2,6), (4,10), (4,12)$	$(2,6), (2,16), (4,10), (3,7),$
		$(3,9), (4,12), (5,15), (5,13)$
$t = 3$	$(12,13)$	$(16,17)$
$t = 4$	$(6,17), (2,3)$	—
$t = 5$	$(17,16), (3,7)$	—
$t = 6$	$(16,15), (7,8)$	—
$t = 7$	$(8,9)$	—

The two last examples give a hint that possibly *plasmodium fails to well approximate a MST because it avoids branching heavily.*.

6.5 Growing from all points

What will happen if every data point is actively searching for its neighbors? To answer the question, we placed a plasmodium — either on its own oat flake (Fig. 6.12a) or on a piece of filter paper (Fig. 6.13a) — in each point of a given dataset. Agar gel was chosen as a substrate to ease observation of propagating pseudopodia.

The plasmodia propagate from their original points forming a fine network of protoplasmic tubes, similarly to diffusive or excitation wave fronts in reaction–diffusion chemical media. When wave fronts of plasmodia orig-

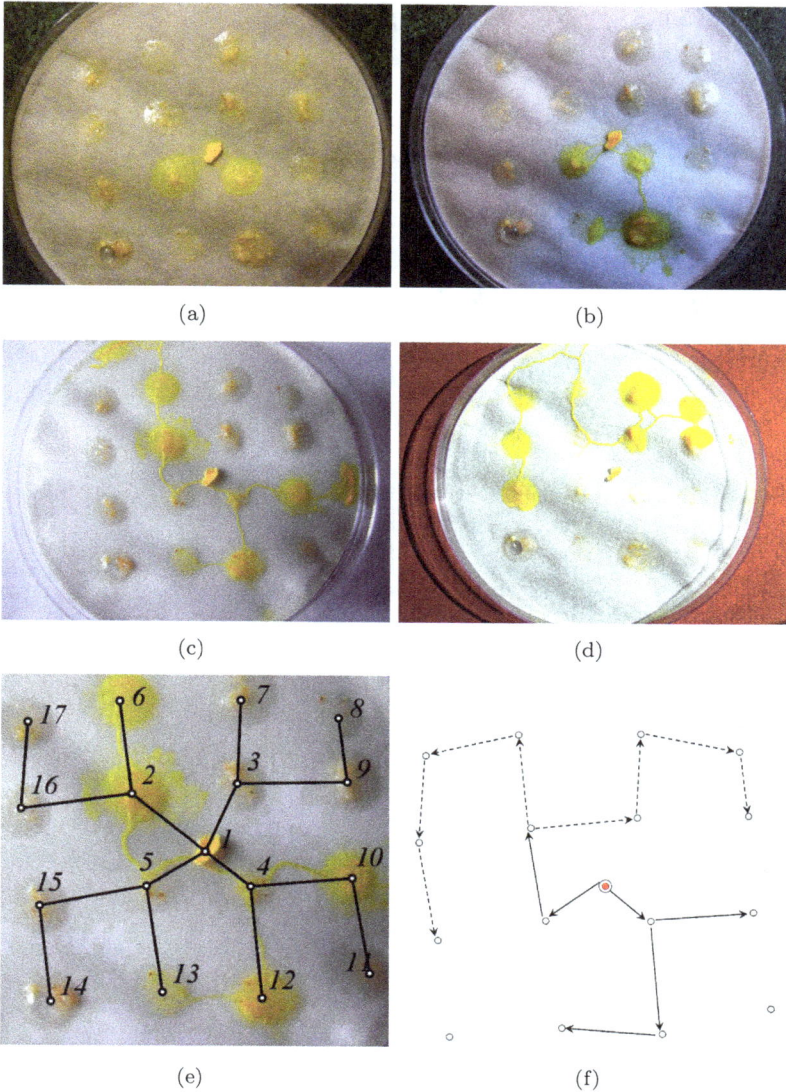

Fig. 6.11 The plasmodium spans almost all data sites in two phases: (a)–(c) first phase, (d) second phase. MST computed on the same data points is shown in (e). Sequence of edge generation by plasmodium is shown in (f): edges constructed in the first phase are solid lines, edges built in the second phase are dashed lines.

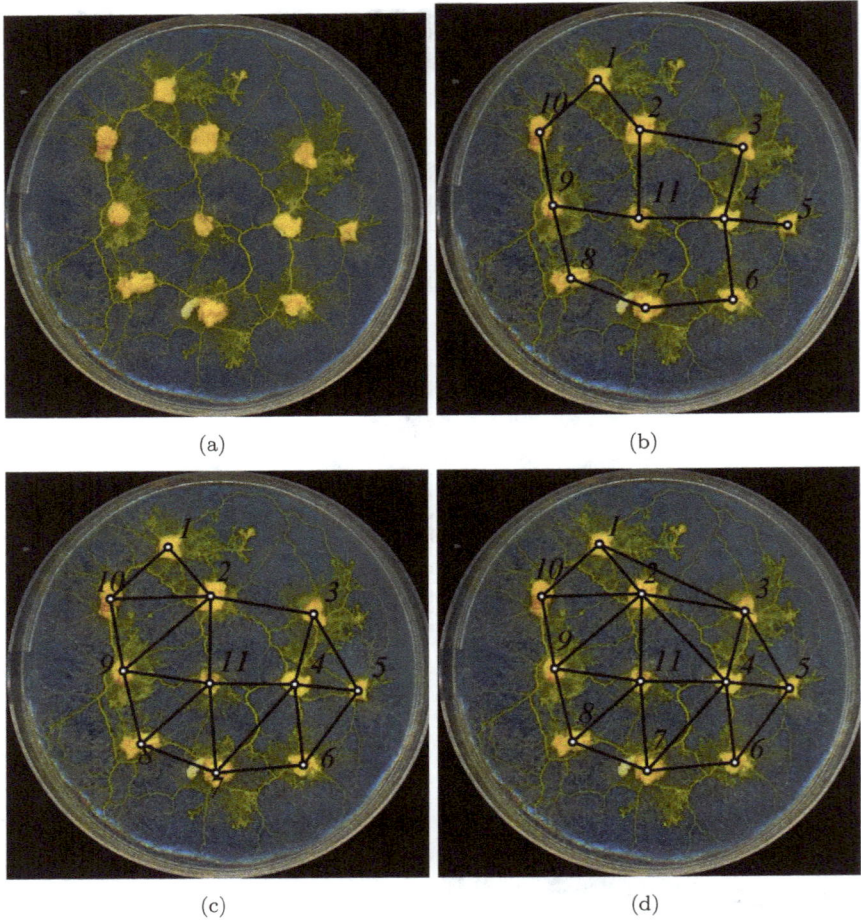

Fig. 6.12 Snapshots of plasmodium protoplasmic network developed in two days, from pieces of plasmodium on oat flakes, placed on 2% agar gel (a). The snapshots are overlaid with proximity graphs computed by standard algorithms: (b) RNG, (c) GG, (d) DT.

inating from two different sources meet, they usually stop their growth. Zones of the substrate where two or more wave fronts 'collide' remain not covered by the fine protoplasmic networks. These zones represent — a very rough — approximation of edges of a Voronoi diagram (see Chap. 3.3).

As demonstrated in Figs. 6.12 and 6.13, the protoplasmic networks developing from all data points simultaneously do not correspond to a RNG. See, for example, the pronounced tube connecting nodes 2 and 4

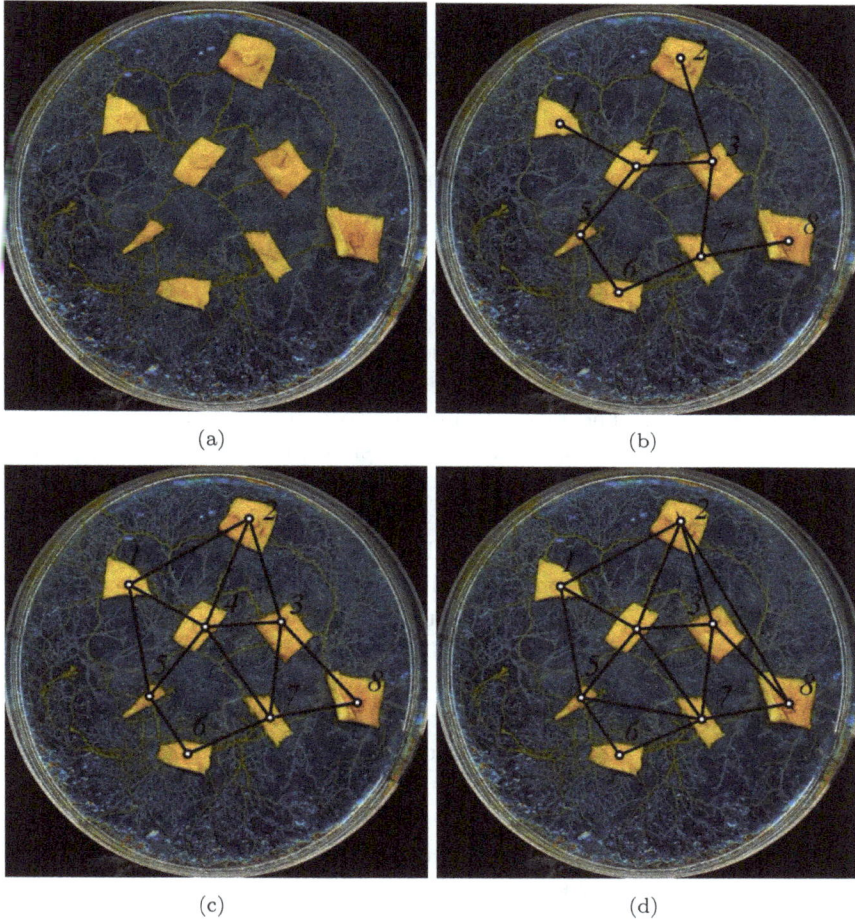

(a) (b)

(c) (d)

Fig. 6.13 Snapshots of plasmodium protoplasmic network developed in three days, from pieces of plasmodium on filter paper, placed on 2% agar gel (a). The snapshots are overlaid with proximity graphs computed by standard algorithms: (b) RNG, (c) GG, (d) DT.

in Fig. 6.12b and the tubes connecting nodes 4 and 2, and nodes 4 and 7, in Fig. 6.13b. This proves that a RNG is not approximated by the plasmodium during simultaneous construction of a protoplasmic proximity graph.

A GG is approximated much more closely if we consider only major protoplasmic tubes connecting the nodes (Figs. 6.12c and 6.13c). The approximation becomes incomplete if secondary protoplasmic tubes are taken into account; see the tube linking nodes 1 and 3 in Fig. 6.12c, and the vein linking nodes 2 and 8 in Fig. 6.13c. These missing links can be found in DTs; see Figs. 6.12d and 6.13d.

Does the plasmodium approximate the exact DT of a given planar set? It does not, as can be seen in Fig. 6.13d: there is an edge (56) in the DT but no protoplasmic tube connecting nodes 5 and 6.

Our experimental results show that the plasmodium of *P. polycephalum* — when it develops in parallel from all points of a given planar set — approximates a RNG with more pronounced protoplasmic tubes and a Delaunay triangulation with additional less pronounced tubes. Neither of the graphs is approximated exactly but rather a graph 'between' a RNG and a DT is developed. The protoplasmic network contains more edges than the corresponding RNG but sometimes fewer edges than the corresponding DT.

This confirms our previous analysis, discussed in [Shirakawa et al. (2009)]. The finding also indicates that a RNG is a primary graph constructed by the plasmodium in parallel implementation, while additional edges, belonging to a DT, are produced rather as auxiliary connections. Thus, we demonstrated that part GG⊆DT is roughly implemented by the plasmodium GG→DT during its development. Because a NNG is a sub-graph of a RNG, and a RNG is a sub-graph of a GG, both a NNG and a RNG are approximated by the plasmodium at the same time as a GG. Thus, a final sequence of graph development will be NNG→RNG→GG→DT.

Returning to the difference in tube thickness, we can speculate that protoplasmic tubes that represent a GG are thicker than auxiliary tubes that represent a DT−GG (i.e. edges existing in a DT and not present in a GG). There may be a biological explanation; see the illustrations in Fig. 6.14. A plasmodium that started its development in point a tries to optimize protoplasmic flow with its neighboring points, where the neighborhood can be satisfactorily expressed in terms of adjacent Voronoi cells. Therefore, it will not be optimal to route a main, i.e. a thicker, tube, which transports protoplasm and nutrients between sites a and b, via a Voronoi neighborhood of site c. This observation corresponds to the original algorithm for pruning a GG from a DT by removing those edges of VD [Matual and Sokal (1984); Jaromczyk and Toussaint (1992)]. Such edges however do sometimes emerge in protoplasmic networks as supplementary transport routes, possibly for the sake of reliability and fault tolerance.

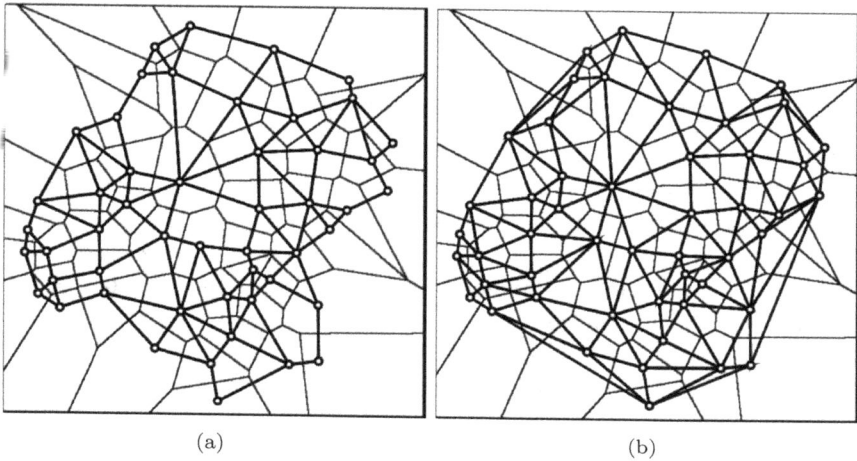

(a)   (b)

Fig. 6.14   Voronoi diagram (shown by thin lines) of 50 planar points overlaid with RNG (a) and GG (b).

6.6   Physarum hierarchy

When inoculated in a single data point the plasmodium builds a NNG, a partial spanning tree (PST) and then a proximity graph (PG), which includes a RNG but not necessarily is included in a GG. If the plasmodium starts growing from each one of given planar points, then the protoplasmic network represents a GG with thick, or main, tubes, and it represents edges of a DT not present in a GG by thin, or auxiliary, tubes.

Finding 6.   *Physarum partially follows the Toussaint hierarchy, when growing proximity graphs in its protoplasmic network: $NNG{\rightarrow}PST{\rightarrow}RNG{\rightarrow}PG$, where a PST is not necessarily a connected graph, $|PST| \geq |MST|$, $RNG{\subseteq}PG$, $PG{\not\subseteq}GG$.*

Finding 7.   *The Physarum follows the upper part $(GG{\rightarrow}DT)$ of the Toussaint hierarchy when growing from all data points simultaneously.*

6.7   Summary

We provided results of scoping experiments on developing proximity graphs in plasmodium of *P. polycephalum*. We partly verified our hypothesis that the protoplasmic graphs developed by the plasmodium in some degree follow the Toussaint hierarchy of proximity graph inclusion.

6.7.1 *Complexity of plasmodium computation*

If we are to measure a time complexity of the plasmodium graph constructed in a number of hops between two neighboring points/nodes, then
the plasmodium approximates a proximity graph in $O(n)$ time steps, and
$\phi(logn)$ on average. The boundaries are good; however, there is a catch.
Pre-processing is done by a human operator during positioning of the oat
flakes. The Physarum machine gets an already geographically ordered set
of planar points. Each pseudopodium then propagates, due to positive
chemo-taxis, along gradients (Fig. 6.2a) and locates nearest sources of nutrients.

6.7.2 *Halting problem*

The plasmodium never stops modifying its protoplasmic network. To get
an answer of when exactly any particular graph is computed, we must
continuously observe the plasmodium's behavior. The Physarum machine
never gives a precise answer to our questions; it rather produces a series
of answers. In this sense the plasmodium computer is akin to Burgin's
inductive Turing machine, it "... does not need to stop to produce a result
of a computation" [Burgin (1983, 1988)]. If we still want to follow traits
of traditional computer science, we can measure resistance between two
arbitrary data points, and thus determine if all planar points are connected
by the protoplasmic graph.

6.7.3 *Reusability*

When dealing with computing in spatially extended nonlinear chemical
media [Adamatzky et al. (2005)], we outlined two types of computers:
reusable without volatile memory, e.g. Belousov–Zhabotinsky reactors, and
disposable (not reusable) with non-volatile memory, e.g. precipitating palladium processors.

Proposition 6.1. *The plasmodium is a reusable non-volatile-memory processor.*

    When the computation is finished, one reduces the humidity, thus caus-
ing the plasmodium to migrate to a certain spot and form there a compact

hard body of cytoplasm — a sclerotium. While migrating, the plasmodium leaves a carcass of 'empty' (but still visible by the naked eye) protoplasmic tubes behind. The discarded protoplasmic tubes represent a result of computation 'written' in a long-term memory of the Physarum machine. The sclerotium can be recultivated in a humid environment to tackle a new computational task.

6.7.4 *Further studies*

Real-life implementations are always far more unclear than results of theoretical studies. The plasmodium does not compute the proximity graphs perfectly. When providing illustrations for the paper, we have indeed manually picked examples of the most successful experiments on developing proximity graphs by Physarum machines. In a few experiments we observed that in the same experimental container, one part of the plasmodium computes one type of proximity graph while another part approximates a different type of graph.

The unpredictability of Physarum machine behavior in no way diminishes its computational potential. More experiments are required on controllability of the foraging behavior and synchronization of the plasmodium's parts when implementing computational tasks. The control of the plasmodium development could be implemented using temperature, light and electromagnetic fields.

Another subject which needs to be cleared up is how close the known proximity graphs are approximated by the plasmodium. And, what are the statistical characteristics of the new graphs, PST and PG, constructed during plasmodium development. We need to identify automatically proximity graphs from protoplasmic networks. The problem of identification remains almost virgin, and is still actually in the graph-theoretic domain, over 20 years after its first formulation in [Toussaint (1989)].

It would also be interesting to consider a few other families of proximity graphs, including β-skeleton (for β changing from 1 to 2) [Kirkpatrick and Radke (1985)] and Urquhart graphs [Urquhart (1980)]. And, possibly, new types of biology-inspired proximity graphs will be discovered on the way.

Chapter 7

Physarum gates

A device is called computationally universal if it implements a functionally complete system of logical gates, e.g. a tuple of negation and conjunction, in its space–time dynamics. Given a new material or a biological substrate, to show that it can be a general-purpose computer one needs to demonstrate computational universality of the material or the substrate. The easiest way to do it is to constrain the substrate into channels and allow disturbances to propagate along the channels and interact with other disturbances at the junctions between the channels.

Such an approach is proved to be fruitful in reaction–diffusion chemical computing. The first ever designs of logical gates in a geometrically constrained Belousov–Zhabotinsky medium were outlined in [Steinbock et al. (1996)]. Elementary gates, like OR and XNOR, and cascades of gates, like AND(OR($\cdot,\cdot$),OR($\cdot,\cdot$)), are executed in laboratory experiments with and numerical simulations of a geometrically constrained BZ medium [Steinbock et al. (1996)]. Other implementations of computing schemes in a geometrically constrained Belousov–Zhabotinsky medium include logical gates for Boolean and multiple-valued logics [Sielewiesiuk and Górecki (2001); Motoike and Yoshikawa (2003); Górecki et al. (2009); Yoshikawa et al. (2009)], many-input logical gates [Górecki and Górecka (2006,a)] and counters [Górecki et al. (2003)]. The only non-excitable chemical implementation of logical gates is the XOR gate device realized in a palladium precipitating processor [Adamatzky and De Lacy Costello (2002)].

What types of Boolean gates can be implemented with plasmodium of *Physarum polycephalum*? How are these gates realized? Can Physarum-based gates be cascaded into advanced computing circuits? The answers are given in this chapter.

7.1   XOR **gate anyone?**

Physarum computes a Voronoi diagram very similarly to precipitating chemical processors (see Chap. 3.3). Precipitating processors [Adamatzky and De Lacy Costello (2002)] compute an XOR gate , so Physarum should.

In [Adamatzky and De Lacy Costello   (2002)], we experimentally demonstrated that a palladium reaction–diffusion processor, geometrically constrained to a T-shaped substrate, implements an XOR gate in its development. The logical gate was designed as follows. An agar gel is mixed with palladium chloride. A T-shape is cut from the gel (Fig. 7.1a). Two horizontal channels, shoulders, are assigned logical inputs x and y. A vertical channel is assumed to be the logical output x. The TRUE values of logical variables are represented by application of potassium iodide, placed at the proximal parts of x or y channels.

As an uncolored solution of the potassium iodide diffuses into the palladium chloride gel, it reacts with the palladium chloride to form dark-colored iodo-palladium species. If potassium iodide is applied only in one of the input channels, the x input in the example of Fig. 7.1b–d, the precipitate is formed in all other channels of the gate, including the z-output channel (Fig. 7.1d). Thus, we get the transformations $\langle 0, 0 \rangle \to 0$, $\langle 0, 1 \rangle \to 1$ and $\langle 1, 0 \rangle \to 1$.

To represent inputs $x = 1$ and $y = 1$, we apply palladium chloride in x and y channels at the same distance from the T-junction (Fig. 7.1e). The gel substrate is uniform; therefore, diffusing fronts of potassium iodide reach the junction at the same time. Substrate loci where two diffusive waves meet remain uncolored (Fig. 7.1f). See [De Lacy Costello (2003); De Lacy Costello and Adamatzky (2003); De Lacy Costello et al. (2004,a, 2009)] for an answer to the question "why the loci remain uncolored?". Absence of iodo-palladium species in the z channel represents logical TRUE, presence — logical FALSE. Thus, we get the transformation $\langle 1, 1 \rangle \to 0$.

If we reproduce experiments analogous to a precipitating XOR gate with Physarum, we do not obtain analogous results with inputs $x = 1$ and $y = 1$. Input to output transformations $\langle 0, 0 \rangle \to 0$, $\langle 0, 1 \rangle \to 1$ and $\langle 1, 0 \rangle \to 1$ are obvious. If there is no plasmodium in the inputs, it does not appear in the output. A plasmodium inoculated in one of the input channels colonizes the whole T-shaped substrate.

If plasmodia are placed in both input channels, they propagate towards the T-junction (Fig. 7.2a). The plasmodia merge immediately after colliding and propagate down the output channel as a single plasmodium

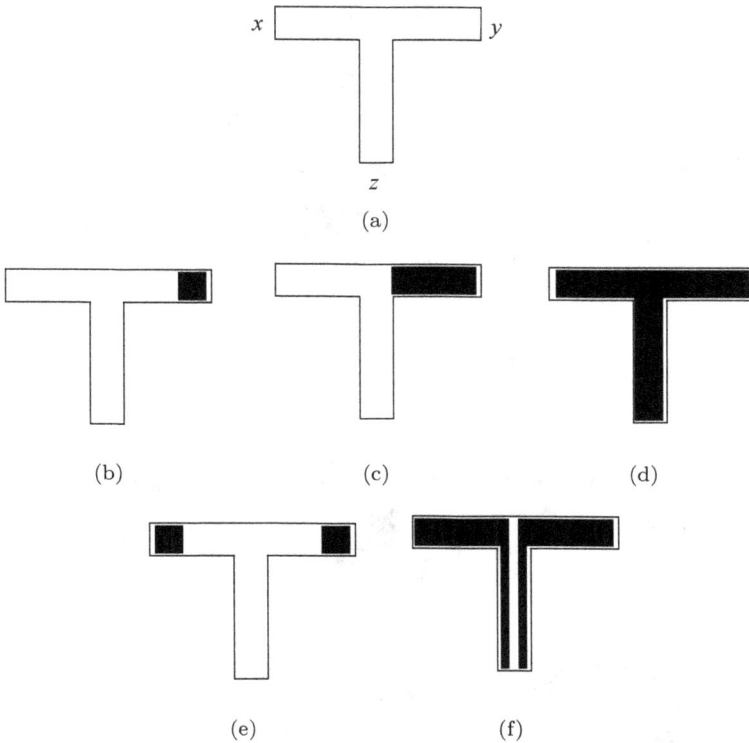

Fig. 7.1   Scheme of palladium gate functioning: (a) structure of the gate, (b)–(d) inputs $x = 0$ and $y = 1$, (e) and (f) inputs $x = 1$ and $y = 1$, (b) initial application of reactant in x-input channel, (cd) precipitation in the gate, (e) initial application of reactant in x- and y-input channels simultaneously, (f) formation of uncolored bisector, which represents the output $z = 0$.

(Fig. 7.2b). Thus, we obtained experimental evidence of the transformation $\langle 1, 1 \rangle \to 1$. This is an AND gate, indeed. But, an AND gate is easy to implement in almost any system.

Finding 8.   *Plasmodium does not implement an* XOR *gate in a T-junction.*

We could not construct anything useful with just an AND gate and not even negation is in hand. Let us consider other approaches then.

In 2004, Tsuda, Aono and Gunji [Tsuda et al.   (2004)] demonstrated in laboratory experiments a realization of Boolean logic negation and conjunction by plasmodium of *P. polycephalum*. In 2004, Adamatzky and De Lacy Costello established in numerical simulation [Adamatzky (2004)] and

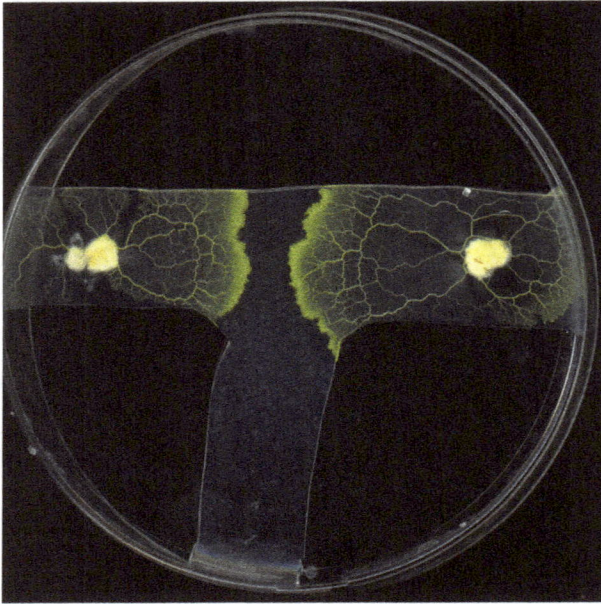

(a) $t = 6$ h

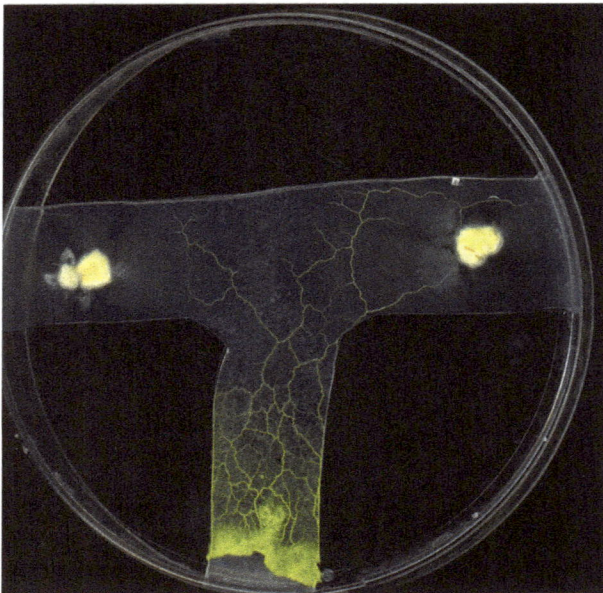

(b) $t = 7$ h

Fig. 7.2 An attempt to reproduce an XOR gate with Physarum: (a) initial propagation of plasmodia, which represent inputs $x = 1$ and $y = 1$, (b) output of the gate.

chemical laboratory experiments [De Lacy Costello and Adamatzky (2005)] that by colliding localized excitations, or wave fragments, in an excitable chemical medium one can implement a functionally complete set of logical gates. We merge the approaches [Tsuda et al. (2004)] and [Adamatzky (2004); De Lacy Costello and Adamatzky (2005)] in the present chapter. We adapt concepts of collision-based computing [Adamatzky (2003)] to realms of Physarum behavior, and develop experimental prototypes of two-input two-output Boolean logic gates.

* * *

Further, we discuss results of experiments with plasmodium propagating on a geometrically constrained substrate: channels and junctions physically representing logical gates are cut from non-nutrient 2% agar plates (Select Agar, Sigma Aldrich).

7.2 Ballistics of Physarum localizations

Proposition 7.1. *Given a cross junction of agar channels and plasmodium inoculated in one of the channels, the plasmodium propagates straight through the junction.*

We experimentally found that plasmodium propagates under its own momentum when no gradients of repellents or attractants are applied. In the example shown in Fig. 7.3a, the plasmodium is inoculated in the northern-most part of the north–south channel. The plasmodium has the only option — to propagate south — because there is no gel substrate further north. Thus, an internal 'momentum' is formed. No food sources are applied on the substrate to attract the plasmodium. The plasmodium propagates straight forward by itself. It does not branch at the junction and moves until it reaches the southern end of the north–south channel (Fig. 7.3a).

In 21 out of 28 trials the plasmodium exhibits clear 'ballistic' behavior and propagates straight through the junction as under its own momentum. In seven out of 28 trials the plasmodium turns into other channels or branches into several channels at once. Adding sources of attractants — assuming that they are balanced over the channels (i.e. if there is an attractant in the east–west channel, there should be one in the north–south

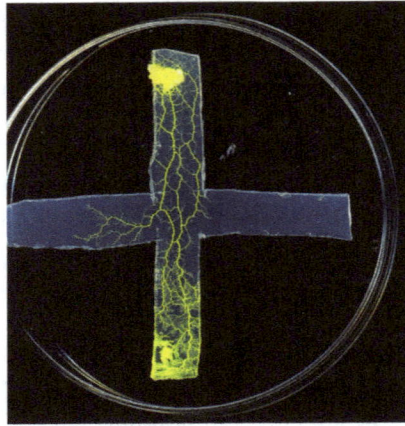

(a)

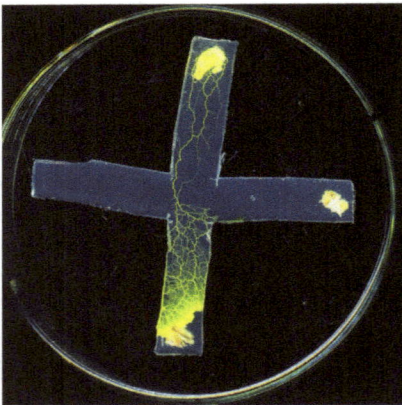

(b)

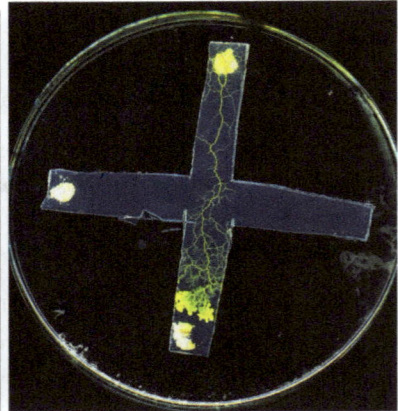

(c)

Fig. 7.3 Experimental examples of plasmodia moving under their own momenta: (a) no sources of chemo-attractants, (b) and (c) oat flakes are placed in southern and eastern (b) and southern and western (c) ends of the channels.

channel) — slightly speeds up propagation of plasmodium but does not change the overall statistics of the plasmodium propagation. Examples are shown in Fig. 7.3bc. In the experiments described in Chap. 7.3, we always used oat flakes to stimulate the plasmodium growth.

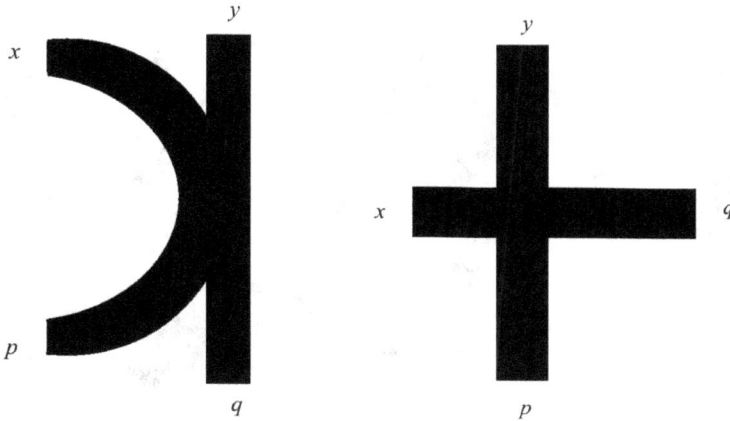

Fig. 7.4 Geometrical structure of Physarum gates P_1 (a) and P_2 (b): x and y are inputs, p and q are outputs.

7.3 Physarum gates

The geometrical structure of gates P_1 and P_2 is shown in Fig. 7.4. We experimented with various shapes of agar and found that the most suitable templates are those shown in Fig. 7.4. Input variables are x and y and outputs are p and q. The presence of a plasmodium in a given channel indicates TRUTH and absence — FALSE.

Each gate implements the transformation $\langle x, y \rangle \rightarrow \langle p, q \rangle$. Experimental examples of the transformations are shown in Fig. 7.5. Plasmodium inoculated in input y propagates along the channel yq and appears in the output q (Fig. 7.5a). Plasmodium inoculated in input x propagates until the junction of x and y, 'collides' with the impassable edge of the channel yq and appears in the output q (Fig. 7.5b).

When plasmodia are inoculated in both inputs x and y, they appear in both outputs p and q (Fig. 7.5c–e). In some cases plasmodia originated in different inputs avoid each other (Fig. 7.5c and d) and thus head towards different outputs. In other cases the plasmodia merge into a single plasmodium but nevertheless branch towards different outputs (Fig. 7.5e). There are no strict rules on repelling and merging and often initial repelling between two plasmodia can be followed by merging (Fig. 7.5f and g).

Typically for biological substrates, plasmodium of *Physarum polycephalum* is sensitive to experimental conditions. The plasmodium sometimes deviates from the general scenario when implementing the transfor-

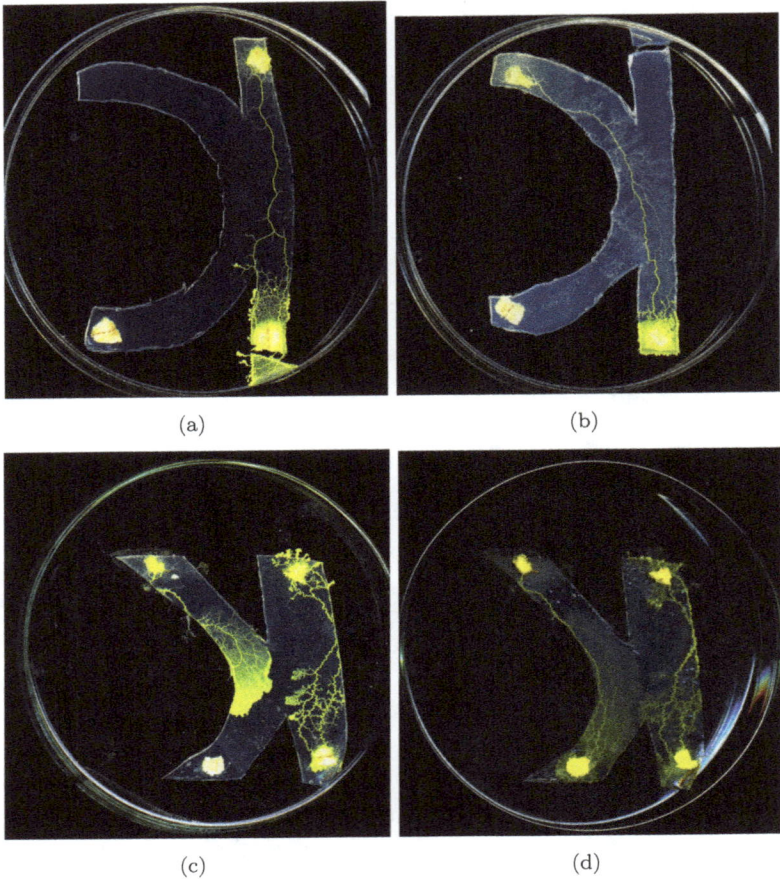

Fig. 7.5 Experimental examples of the transformation $\langle x, y \rangle \to \langle p, q \rangle$ implemented by Physarum gate P_1. (a) $\langle 0, 1 \rangle \to \langle 0, 1 \rangle$, (b) $\langle 1, 0 \rangle \to \langle 0, 1 \rangle$, (c)–(e) $\langle 1, 1 \rangle \to \langle 1, 1 \rangle$. (f) and (g) snapshots of P_1 gate taken at 12 h interval.

mation $\langle x, y \rangle \to \langle p, q \rangle$. Results of experiments with Physarum gate P_1 are shown in Fig. 7.6a.

Plasmodia do not cancel each other at once. Therefore, if at least one of the inputs is '1', we expect to see '1' at one of the outputs. Input scenario $\langle 1, 1 \rangle$ is straightforward: in six out of seven experiments plasmodia appear on both outputs. Thus, we obtain the transformation $\langle 1, 1 \rangle \to \langle 1, 1 \rangle$ (Fig. 7.6a).

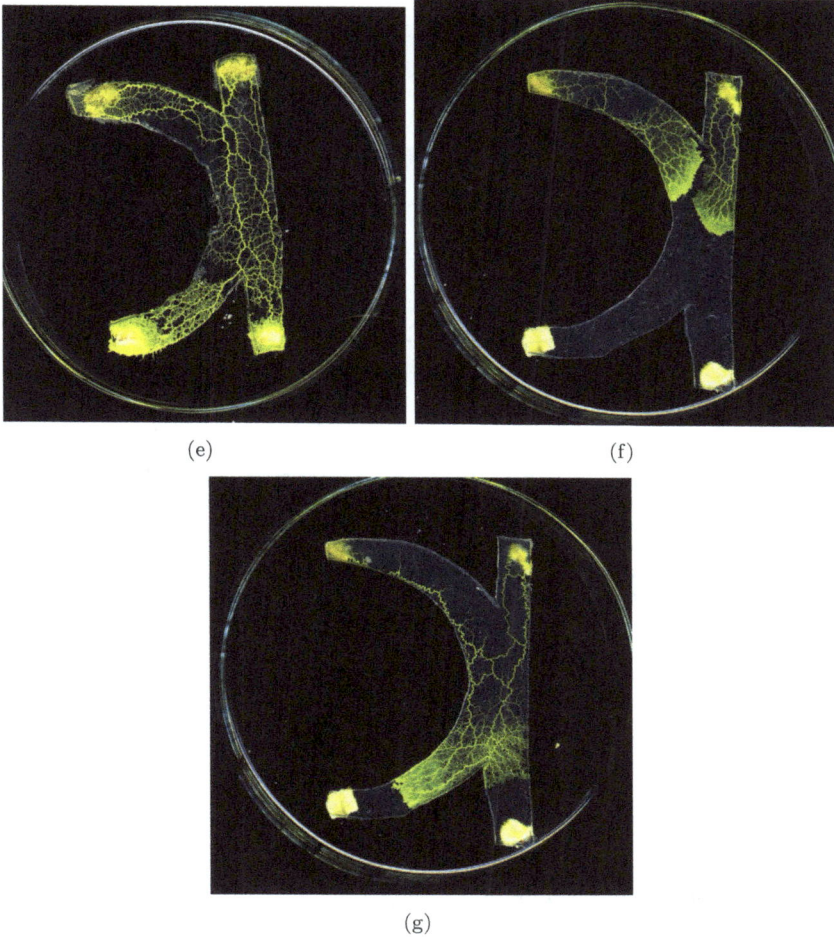

(e)

(f)

(g)

Fig. 7.5 *Continued.*

Plasmodium inoculated in input y (while input x is empty) will appear in output q in 17 out of 22 experiments. Thus, the transformation $\langle 0, 1 \rangle \rightarrow \langle 0, 1 \rangle$ is realized by plasmodium in over 70% of cases.

The input combination $\langle 1, 0 \rangle$ gives us less stable results: in nine out of 13 experiments (69%) the plasmodium reaches output q. The plasmodium enters output p in three out of 13 experiments, and the plasmodium branches inside both outputs in one experiment. Nevertheless, the trans-

x	y	p	q	Frequency	x	y	p	q	Frequency
0	0	0	0	0	0	0	0	0	0
0	1	0	0	0	0	1	0	0	0
		0	1	$\frac{17}{22}$			0	1	$\frac{4}{29}$
		1	0	$\frac{2}{22}$			1	0	$\frac{21}{29}$
		1	1	$\frac{3}{22}$			1	1	$\frac{4}{29}$
1	0	0	0	0	1	0	0	0	0
		0	1	$\frac{9}{13}$			0	1	$\frac{16}{27}$
		1	0	$\frac{3}{13}$			1	0	$\frac{5}{27}$
		1	1	$\frac{1}{13}$			1	1	$\frac{6}{27}$
1	1	0	0	0	1	1	0	0	0
		0	1	$\frac{1}{7}$			0	1	$\frac{13}{21}$
		1	0	0			1	0	$\frac{4}{21}$
		1	1	$\frac{6}{7}$			1	1	$\frac{5}{21}$
(a)					(b)				

Fig. 7.6 Experimental data on the transformations $\langle x, y \rangle \rightarrow \langle p, q \rangle$ implemented by Physarum gates (a) P_1 and (b) P_2. Values $x = 1$ and $y = 1$ are represented by plasmodia inoculated in inputs x and y, respectively. Values $p = 1$ and $q = 1$ are represented by plasmodia reaching outputs p and q, respectively. The frequency of each particular scenario $\langle a, b \rangle \rightarrow \langle c, d \rangle$ is presented by the fraction: the denominator is the total number of experiments for $\langle x, y \rangle = \langle a, b \rangle$ and the numerator is the number of experiments completed with the output tuple $\langle x, y \rangle = \langle c, d \rangle$.

formation $\langle 1, 0 \rangle \rightarrow \langle 0, 1 \rangle$ is realized by plasmodium in well over half of the trials (Fig. 7.6a).

Finding 9. *Plasmodium of* Physarum polycephalum *implements a two-input two-output Boolean gate* $\langle x, y \rangle \rightarrow \langle xy, x + y \rangle$ *with reliability exceeding 69%.*

Experimental snapshots of plasmodium propagating in the gate P_2 are shown in Fig. 7.7. Taking input x as empty, plasmodium placed in input y usually (see statistics in Fig. 7.6) propagates directly towards output q (Fig. 7.7a and b). Plasmodium inoculated in input x (when input y is empty) travels directly towards output p (Fig. 7.7c). Thus, the transformations $\langle 0, 1 \rangle \rightarrow \langle 1, 0 \rangle$ and $\langle 1, 0 \rangle \rightarrow \langle 0, 1 \rangle$ are implemented.

The gate's structure is asymmetric; the x channel is shorter than the y channel. Therefore, the plasmodium placed in input x usually passes the junction by the time plasmodium originated in input y arrives at the junction (Fig. 7.7d–f). The y plasmodium merges with the x plasmodium

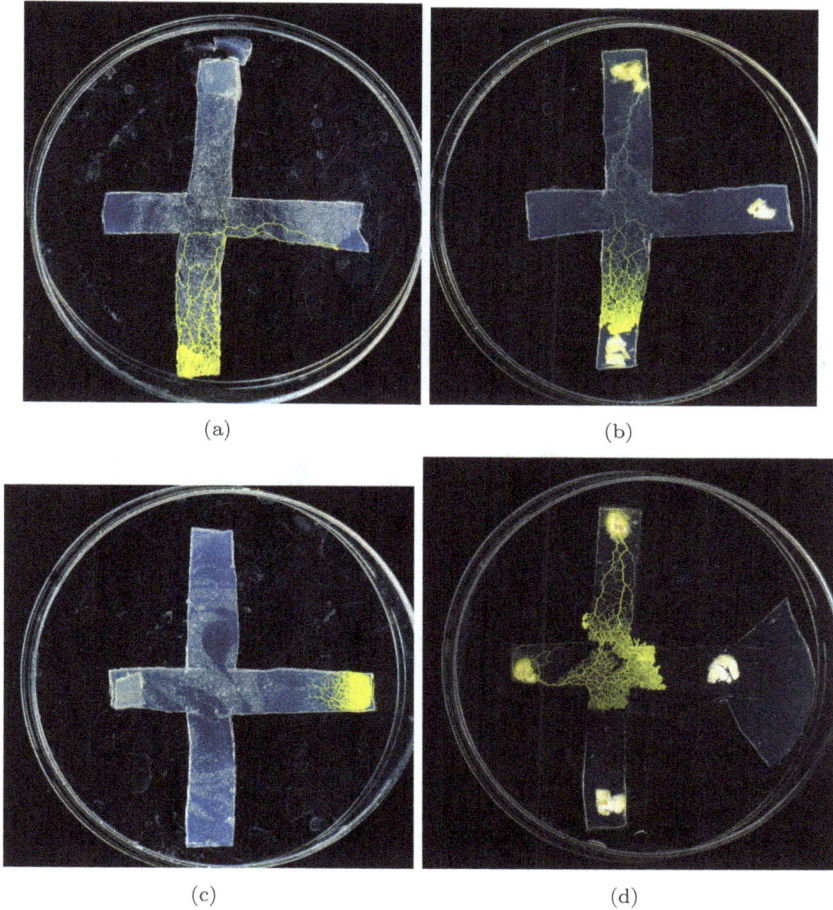

(a) (b)

(c) (d)

Fig. 7.7   Experimental examples of the transformation $\langle x, y \rangle \to \langle p, q \rangle$ implemented by Physarum gate P_2: (a) $\langle 0, 1 \rangle \to \langle 1, 0 \rangle$, plasmodium is inoculated in input y, no oat flakes present; (b) $\langle 0, 1 \rangle \to \langle 1, 0 \rangle$, oat flakes are placed in both outputs; (c) $\langle 1, 0 \rangle \to \langle 0, 1 \rangle$, no oat flakes present; (d) and (e) two snapshots of the transformation $\langle 1, 1 \rangle \to \langle 0, 1 \rangle$, taken with 11 h interval, oat flakes are placed in both outputs; (f) $\langle 1, 1 \rangle \to \langle 0, 1 \rangle$, oat flakes are placed in both outputs; (g) and (h) the transformations $\langle 1, 1 \rangle \to \langle 0, 1 \rangle$ are less pronounced than in previous examples; however, we see that output q is more extensively occupied by plasmodium than output p.

and they both propagate towards output q. Extension of the gel substrate after output q does usually facilitate implementation of the transformation $\langle 1, 1 \rangle \to \langle 0, 1 \rangle$ (Fig. 7.7d–h).

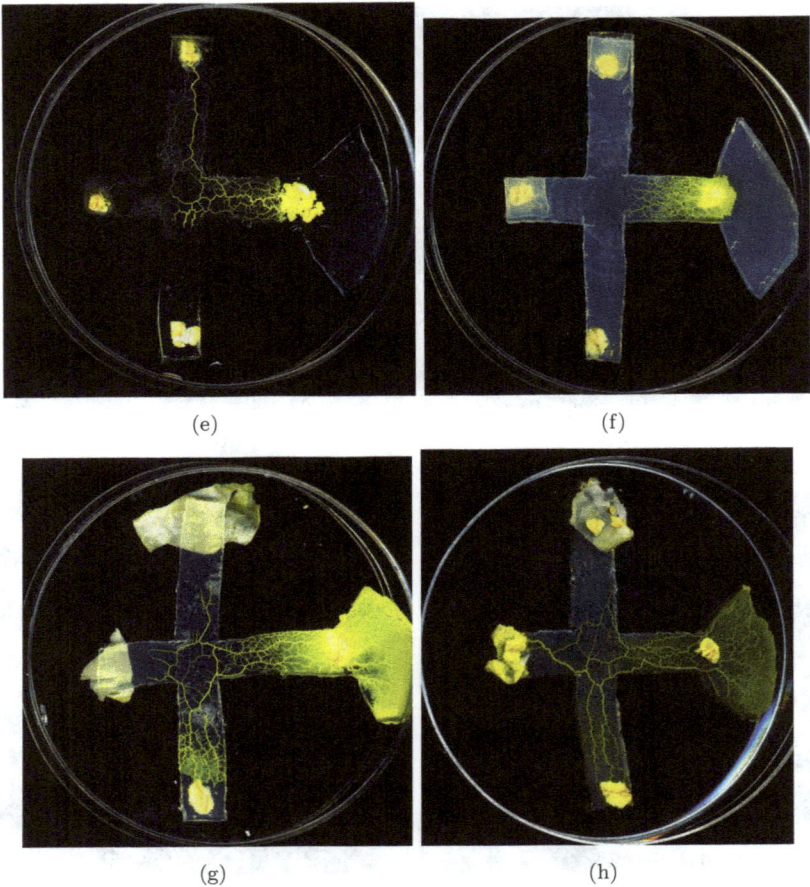

(e)   (f)

(g)   (h)

Fig. 7.7   *Continued.*

Frequencies of various input–output transformations occurring in experiments are shown in Fig. 7.6b. Plasmodium inoculated in input y will reach only output p in 21 out of 29 experiments. The transformation $\langle 1, 0 \rangle \rightarrow \langle 0, 1 \rangle$ takes place in 16 out of 27 experiments. The transformation $\langle 1, 1 \rangle \rightarrow \langle 0, 1 \rangle$ occurs in 13 out of 21 experiments.

Finding 10. *A Physarum machine implements a two-input two-output gate* $\langle x, y \rangle \rightarrow \langle x, \overline{x}y \rangle$ *with reliability exceeding 59%.*

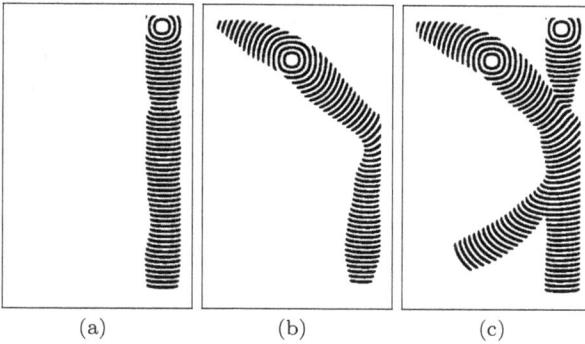

Fig. 7.8 Time-lapse images of localized excitations, wave fragments, traveling in channels of gate P_1 filled with excitable medium. Dynamics of the excitable medium gate P_1 are shown during implementation of the transformations (a) $\langle 0, 1 \rangle \rightarrow \langle 0, 1 \rangle$, (b) $\langle 1, 0 \rangle \rightarrow \langle 0, 1 \rangle$, (c) $\langle 1, 1 \rangle \rightarrow \langle 1, 1 \rangle$. The transformation $\langle 1, 1 \rangle \rightarrow \langle 1, 1 \rangle$ is simulated for initial excitations in channels x and y positioned at equal distances from their meeting point.

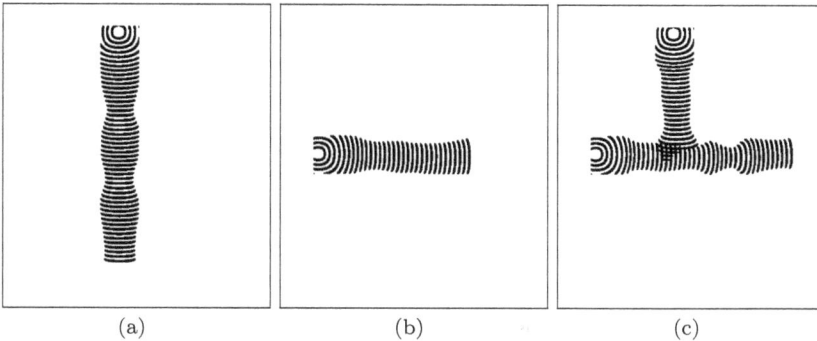

Fig. 7.9 Time-lapse images of localized excitations, wave fragments, traveling in channels of gate P_2 filled with excitable medium. Dynamics of the excitable medium gate P_2 are shown during implementation of the transformations (a) $\langle 0, 1 \rangle \rightarrow \langle 1, 0 \rangle$, (b) $\langle 1, 0 \rangle \rightarrow \langle 0, 1 \rangle$, (c) $\langle 1, 1 \rangle \rightarrow \langle 0, 1 \rangle$.

7.4 Simulation of Physarum gates

Experimental Findings 1 and 2 are confirmed in numerical simulation using a two-variable Oregonator model of a sub-excitable medium. To represent the value '1' in input channel x or y, we generate an excitation near the entrance of the channel x or y, respectively. Two wave fragments are formed. One travels outside the gate, the other travels towards the outputs.

Wave fragments in sub-excitable media are notably unstable; they keep their shape only for a short period of time. Then the fragments either shrink and annihilate or expand unlimitedly. During simulation of the gates P_1 and P_2, we manually adjusted the parameter ϵ to keep wave fragments from collapsing and expanding. Figures 7.10 and 7.11 illustrate dynamics of ϵ during simulations of the gates P_1 (Fig. 7.8) and P_2 (Fig. 7.9).

Figures 7.10 and 7.11 show how the parameter ϵ (see description of the model in Chap. 4.3) is changed during simulation of the gates P_1 and P_2 in a sub-excitable medium. We also show dynamics of the medium's activity. At each step of the simulation we calculate the activity $\alpha^t = |\mathbf{L}|^{-1} \cdot \sum_{x \in \mathbf{L}} u_x^t$ as a sum of values u for each node of the grid $\mathbf{L}$ normalized by the total number of nodes $|\mathbf{L}|$.

Let us look at the time-lapse images of wave fragments propagating in gate P_1 (Fig. 7.8) and gate P_2 (Fig. 7.9). Scenarios where only one input is excited are straightforward. When $y = 1$ in gate P_1 is initialized, the wave fragment propagates southward along the channel yq (Fig. 7.8a). The wave fragment initiated in input x ($x = 1$) of gate P_1 propagates along the input channel x and collides with the boundary of channel yq (Fig. 7.8b). The wave fragment recovers after collision with the boundary, travels along the channel yq and appears in the output q (Fig. 7.8b). Wave fragments behave similarly in situations of input tuples $\langle 1, 0 \rangle$ and $\langle 0, 1 \rangle$ in gate P_2. They propagate straight along their original input channel and reach outputs opposite to their entry points (Fig. 7.9a and b).

In input scenarios $\langle 0, 1 \rangle$ and $\langle 1, 0 \rangle$, the size of the propagating wave fragment was not enough for the fragment to branch into output channel p of gate P_1 (Fig. 7.8a and b). When two wave fragments are initiated, $x = 1$ and $y = 1$, they collide at the junction of input channels x and y. The wave fragments merge into a single larger wave fragment. This new wave fragment propagates towards output q and also expands into output channel p (Fig. 7.8c).

In gate P_2, for input values $x = 1$ and $y = 1$, a wave fragment originated in the x input arrives at the junction between channels xp and yq a little earlier than the wave fragment originated in the y input. Therefore, the y wave collides into the refractory tail of the x wave. The y wave annihilates (Fig. 7.9c). Such a development is phenomenologically identical to Physarum gate P_2 (Fig. 7.7d and e) with the only difference that the plasmodium 'wave' originated in input y is not annihilated but merges with plasmodium originated in input x.

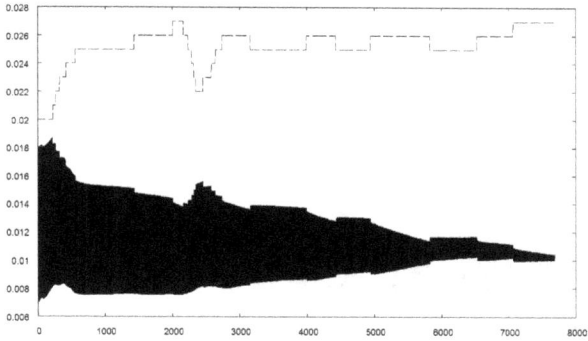

(a)

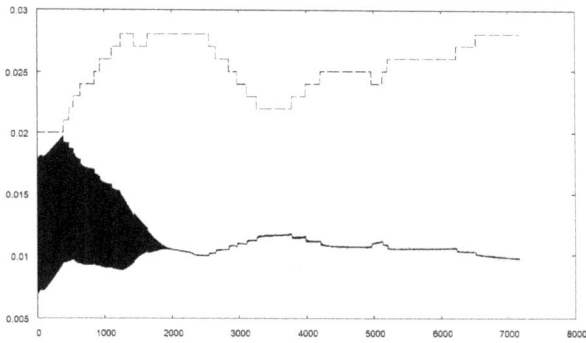

(b)

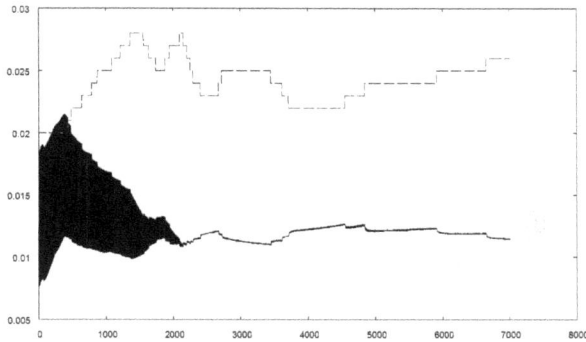

(c)

Fig. 7.10   Dynamics of parameter ϵ, dotted line, and activity α, solid line, during operation cycle of gate P_1 for input tuples (a) $\langle 0, 1 \rangle$ (see Fig. 7.8a), (b) $\langle 1, 0 \rangle$ (see Fig. 7.8b), (c) $\langle 1, 1 \rangle$ (see Fig. 7.8c).

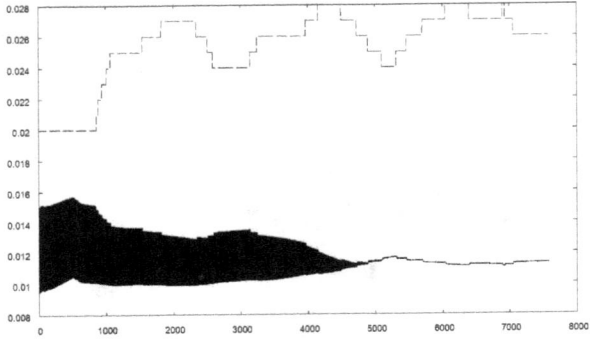

(a)

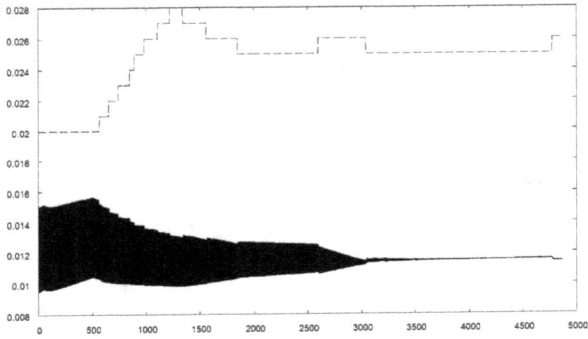

(b)

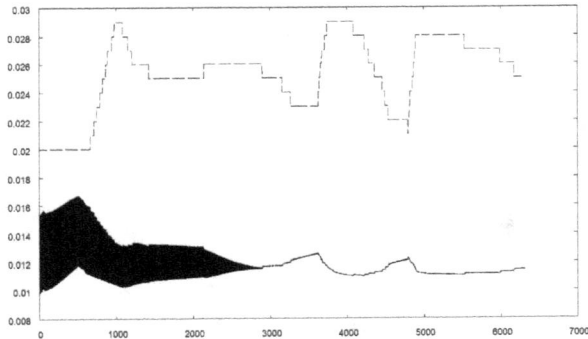

(c)

Fig. 7.11   Dynamics of parameter ϵ, dotted line, and activity α, solid line, during operation cycle of gate P_2 for input tuples (a) $\langle 0, 1 \rangle$ (see Fig. 7.9a), (b) $\langle 1, 0 \rangle$ (see Fig. 7.9b), (c) $\langle 1, 1 \rangle$ (see Fig. 7.9c).

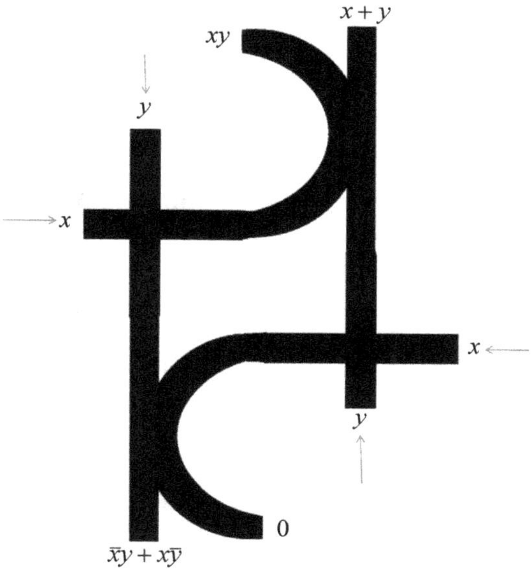

Fig. 7.12   Scheme of one-bit half-adder made of gates P_1 and P_2. Inputs are indicated by arrows. Outputs $\overline{x}y + x\overline{y}$ and xy are sum and carry values computed by the adder. Outputs 0 and $x + y$ are byproducts.

7.5   Simulated one-bit half-adder

A one-bit half-adder is a logical circuit which takes two inputs x and y and produces two outputs: the sum $\overline{x}y + x\overline{y}$ and the carry xy. To construct a one-bit half-adder with Physarum gates, we need two copies of gate P_1 (Fig. 7.4a) and two copies of gate P_2 (Fig. 7.4b). Cascading the gates into the adder is shown in Fig. 7.12. Signals x and y are inputted in P_2 gates. Outputs of P_2 gates are connected to inputs of P_1 gates.

We did not manage to realize a one-bit half-adder in experiments with living plasmodium because the plasmodium behaved differently in the assembly of the gates than in isolated gates. Simulation of the adder using the Oregonator model was successful (Fig. 7.13).

To simulate input values $x = 0$ and $y = 1$, we initiate wave fragments at the beginning of channels marked by y and an arrow in Fig. 7.12. The wave fragments propagate along their channels. The waves do not branch at the junctions with other channels because we keep the wave fragments localized by varying the parameter ϵ (Fig. 7.13a).

For input values $x = 1$ and $y = 0$, wave fragments are originated in

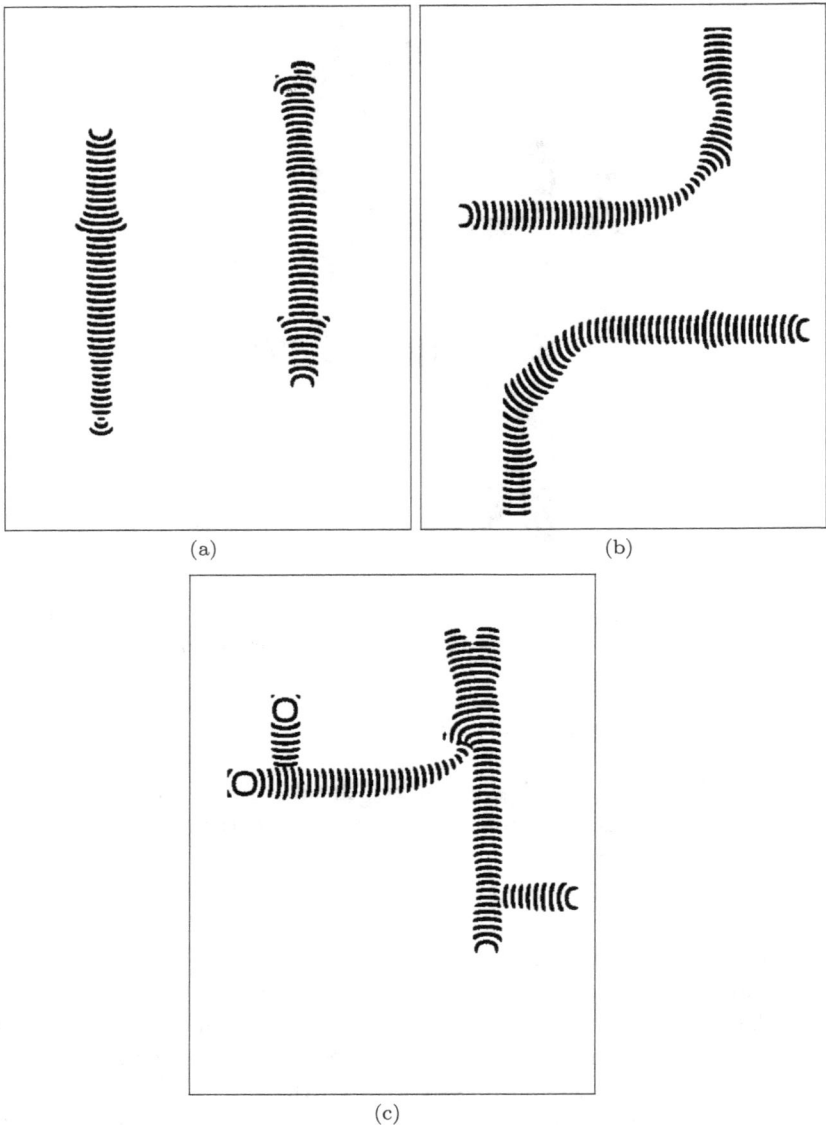

Fig. 7.13   Time-lapse images of localized excitations, wave fragments, traveling in channels of one-bit half-adder filled with excitable medium. Dynamics of excitations are shown for input values (a) $x = 0$ and $y = 1$, (b) $x = 1$ and $y = 0$, (c) $x = 1$ and $y = 1$.

sites marked by x and an arrow in Fig. 7.12. The wave fragment started in the western x-input channel propagates towards the $(x + y)$ output of the adder. The wave fragment originated in the eastern x-input channel travels towards $\overline{x}y + x\overline{y}$ (Fig. 7.13b).

When both inputs are activated, $x = 1$ and $y = 1$, a wave fragment originated in the western y-input channel is blocked by the refractory tail of the wave fragment originated in the western x-input channel. The wave fragment traveling in the eastern x-input channel is blocked by the tail of the wave fragment traveling in the eastern y-input channel. The wave fragments representing $x = 1$ and $y = 1$ enter the top-right gate P_1 and emerge at its outputs xy and $x + y$ (Fig. 7.13b). Thus, functionality of the circuit of Fig. 7.13 is demonstrated.

7.6 Why do we use a non-nutrient substrate?

Our designs are based on the interactions between traveling localizations: plasmodium localizations propagating on a non-nutrient substrate and wave fragments propagating in a sub-excitable medium. Similarities between the plasmodium localizations and wave fragments were discussed in detail [Adamatzky (2009)]. We stress that things go absolutely differently on a nutrient-rich substrate (corn meal agar) and a fully excitable chemical medium. Plasmodium inoculated in any point of the nutrient agar gel gate propagates into all channels (Fig. 7.14a–c). An excitation wave initiated at any point of an excitable medium gate spreads all over the gate (Fig. 7.14d). We can conclude therefore that it is impossible to implement logical functions with plasmodium of *Physarum polycephalum* on a nutrient-rich substrate.

7.7 Summary

We established experimentally and in numerical simulations that plasmodium of *Physarum polycephalum* realizes basic logical operations on a geometrically constrained non-nutrient substrate. We designed two types of Boolean logic gates; both gates have two inputs and two outputs. The gates implement the transformations $\langle x, y \rangle \rightarrow \langle xy, x + y \rangle$ and $\langle x, y \rangle \rightarrow \langle x, \overline{x}y \rangle$. We showed how the Physarum gates can be assembled into a one-bit half-adder. Functionality of the adder was illustrated using the two-variable Oregonator model.

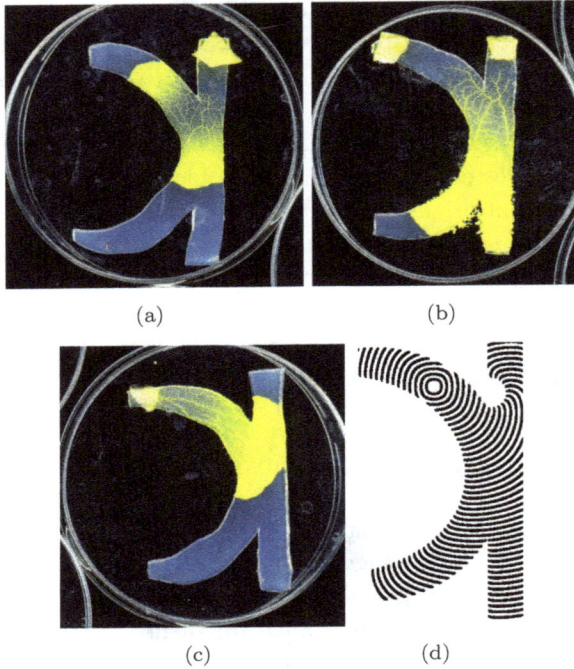

Fig. 7.14 Snapshots of plasmodium propagating in gate P_1 on 2% corn meal agar (a)–(c) and numerical simulation (d): (a) inputs $x = 0$ and $y = 1$, see the gate's scheme in Fig. 7.4, (b) inputs $x = 1$ and $y = 1$, (c) inputs $x = 1$ and $y = 0$, (d) time-lapse images of excitation wave propagating in gate P_1 for inputs $x = 1$ and $y = 0$, simulation of case (c).

The reliability of the experimental Physarum gate is quite low: 69% for gate P_1 and 59% for gate P_2. This is because the behavior of plasmodium is determined by too many environmental factors — thickness of substrate, humidity, diffusion of chemo-attractants in the substrate and in the surrounding air volume, and physiological state of the plasmodium during each particular experiment. Increasing the reliability of Physarum gates might be a scope of further studies.

Chapter 8

Kolmogorov–Uspensky machine in plasmodium

In the early 1950s, Kolmogorov [Kolmogorov (1953)], later with his student Uspensky [Kolmogorov and Uspensky (1958)], outlined a concept of an abstract machine defined on a dynamically changing graph-based structure, called a *Kolmogorov complex*. The Kolmogorov complex is a computational process on a finite undirected connected graph with distinctly labeled nodes. The nodes are labeled in such a manner that any two closest neighbors of any node have different labels. The graph is an analog of a storage structure. A computational process travels on the graph, activates nodes and removes and adds edges. There is only one active node at any step of development, i.e. the neighborhood of any active node is fixed for any particular algorithm [Uspensky (1992)].

The Kolmogorov complex later became known as a *Kolmogorov–Uspensky machine* (KUM). A KUM is defined on a colored/labeled undirected graph with bounded degrees of nodes and bounded number of colors/labels. KUMs operate, modifying their storage, as follows:

(1) select an active node in the storage graph;
(2) specify a local active zone, i.e. the node's neighborhood;
(3) modify the active zone by adding a new node with a pair of edges connecting the new node with the active node;
(4) delete a node with a pair of incident edges;
(5) add/delete the edge between the nodes.

A program for a KUM specifies how to replace the neighborhood of an active node with a new neighborhood, depending on the labels of edges connected to the active node and the labels of the nodes in proximity to the active node [Blass and Gurevich (2003)].

A Turing machine (TM) and a KUM compute the same class of func-

Kolmogorov machines (1953)

Knuths linking automata (1968)

Schönhage storage modification machines (1970s)

Tarjans reference machines (1977)

Random access machines

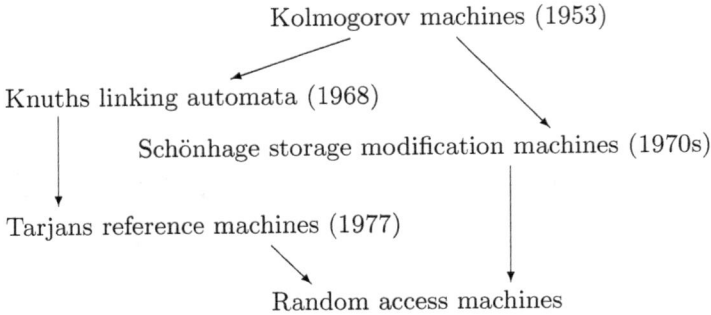

Fig. 8.1   Development of storage modification machines.

tions. Functions computable on Turing machines are computed on KUMs and *vice versa*. Any sequential device can be simulated by a KUM. Essentially any computation performing only one local action at a time is simulated by a KUM [Gurevich (1988)]. However, a KUM is more flexible than a TM because a KUM is not restricted by a fixed topology of storage space. A KUM can be seen as a TM which changes topology of its tape during computation.

Grigoriev (1980) demonstrated that there are some predicates recognizable in real time by a KUM but not recognizable in real time by a TM. As Cloteaux and Ranjan stated [Cloteaux and Rajan (2006)], KUMs are stronger than any model of computation that requires $\Omega(n)$ time to access its memory. They also highlighted results by Shvachko [Shvachko (1991)], who proved that KUMs are stronger than any 'tree' machine — TMs with tapes which are infinite-length binary trees.

A multi-tape TM with space complexity P can be simulated by a storage modification machine (SMM), a successor of a KUM, with $O(P/logP)$ nodes in real time [Boas (1989)]. A SMM is unlikely to be simulated in real time by a KUM [Gurevich (1988)], because a SMM has an unbounded degree of nodes.

KUMs are the mother of all present models of real-life computation. From their inception in the 1950s, they have been reincarnated in 1968 as Knuth's linking automata [Knuth (1968)], in 1977 as Tarjan's reference machines [Tarjan (1977)] and in the 1970s as Schönhage's storage modification machines [Schönhage (1973, 1980)] (Fig. 8.1). This evolution — with its main outcome of relaxing the condition of bounded in- and out-degrees of the storage graph — ultimately reached the state of random access machines, the architectures of modern computers.

Gurevich [Gurevich (1988)] suggested that the edge of the 'Kolmogorov complex' is not only informational but also a physical entity and reflects the physical proximity of the nodes (e.g. even in three-dimensional space the number of neighbors of each node is polynomially bounded).

Turing machines formalize computation as it is performed by a human. Kolmogorov machines formalize computation as it performed by a physical process.[Blass and Gurevich (2003)]

What would be a good natural implementation of a KUM? Reaction–diffusion chemical computers [Adamatzky (2001); Adamatzky et al. (2005)] lack flexibility, and the stationary or dissipative structures formed by them are rather static or quasi-static. DNA and other molecular computers have almost no structure, and act mainly as bulk, well-stirred, media computing devices. Thus, a potential candidate should be capable of growing and unfolding graph-like storage structures, dynamically manipulating nodes and edges, and should have a wide range of functioning parameters. A plasmodium of *P. polycephalum* satisfies all of these requirements.

8.1 Physarum machines

In this section we provide step-by-step comparison of a KUM and plasmodium of *Physarum polycephalum*, and show the following.

Proposition 8.1. *Plasmodium is a biological substrate that implements a Kolmogorov–Uspensky machine.*

8.1.1 *Materials for Physarum machine*

The scoping experiments were designed as follows. We either covered a container's bottom with a piece of wet filter paper and placed a plasmodium on it, or just planted plasmodium on the bottom of a bare container and fixed wet paper to the container's cover to keep humidity high. Oat flakes were distributed throughout the container to supply nutrients and represent data nodes of the Physarum machine.

The containers were stored in the dark except during periods of observation. To color the oat flakes, where required, we used SuperCook Food Colorings[1]: blue (E133, E122), yellow (E102, E110, E124), red (E110, E122)

---

[1]www.supercook.co.uk

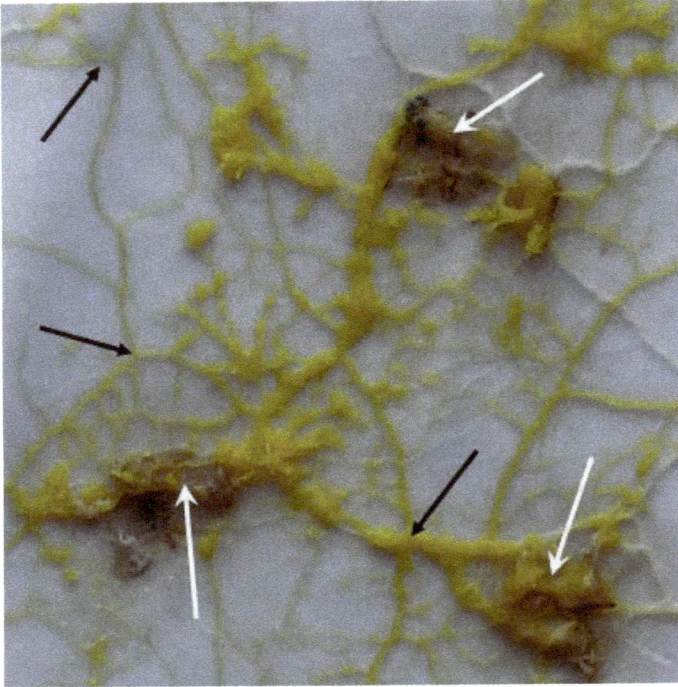

Fig. 8.2 Nodes of a Physarum machine. Stationary nodes are indicated by white arrows; dynamic nodes by black arrows.

and green (E102, E142). Flakes were saturated with the colorings and then dried.

8.1.2 *Nodes*

A Physarum machine has two types of nodes: stationary nodes, represented by sources of nutrients (oat flakes), and dynamic nodes, sites where two or more protoplasmic tubes originate (Fig. 8.2). At the beginning of computation, stationary nodes are distributed in computational space; see the example in Fig. 8.3, and plasmodium is placed at one point in the space. Initially, the plasmodium exhibits foraging behavior, and occupies only stationary nodes (i.e. oat flakes).

(a) 1st day   (b) 2nd day

(c) 3rd day   (d) 4th day

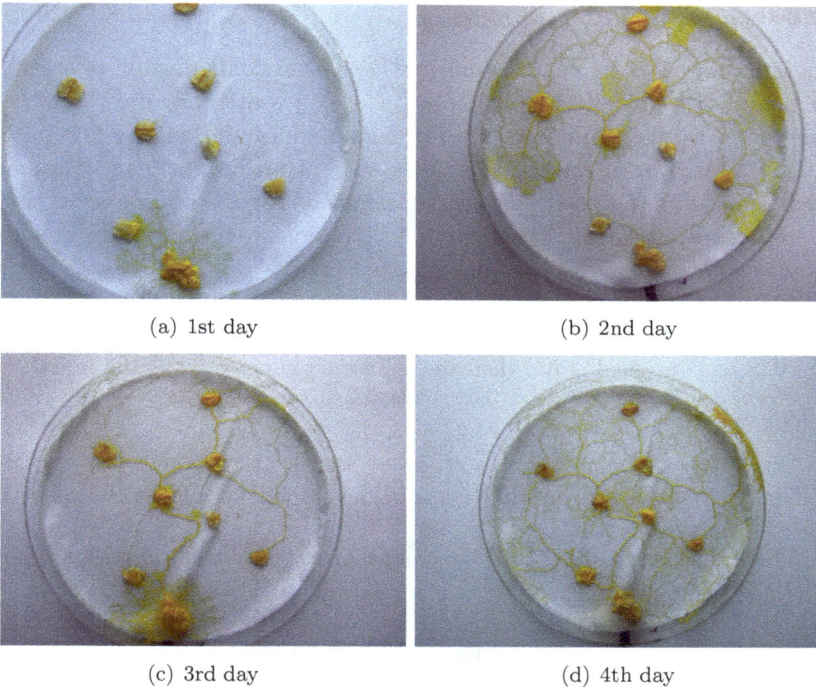

Fig. 8.3   An example of the computational process in a Physarum machine. Photographs (a)–(d) are taken with time lapses of approx. 24 h.

8.1.3   *Edges*

An edge of a Physarum machine is a vein, or a tube, of protoplasm connecting stationary and/or dynamic nodes. A KUM is an undirected graph, i.e. if nodes x and y are connected, then they are connected by two edges (xy) and (yx). In a Physarum machine, this is implemented by a single edge but with periodically reversing flow of protoplasm [Kamiya (1950); Nakagaki et al. (2000)]. The protoplasm stream runs with speed of approx. 1–3 mm/s [Tirosh et al. (1973)], or sometimes up to 4 mm/s [Kamiya (1950)], changing direction every 1–3 min. This may also be associated with periodic inversion of the potential.

8.1.4 *Data, results and halting*

Program and data are represented by the spatial configuration of stationary nodes. Results of the computation over a stationary data node are represented by the configuration of dynamic nodes and edges. The initial state of a Physarum machine includes part of an input string (the part which represents the position of plasmodium relative to stationary nodes), an empty output string, the current instruction in the program and the storage structure consisting of one isolated node.

What is a result of a computation? Kolmogorov and Uspensky wrote:

> "If S is a terminal state, then the connected component of the initial vertex is considered to be the 'solution'" [Kolmogorov and Uspensky (1958)].

That is, the whole graph structure developed by plasmodium is the result of its computation.

When discussing KUMs, Blaas and Gurevich [Blass and Gurevich (2003)] pointed out that

> "the process runs until either the next step is impossible or a signal says the solution has been reached".

Plasmodium proceeds with computation even when the solution has been reached and halts only when physical resources are exhausted. Plasmodium continues its spreading, reconfiguration and development as long as there are enough nutrients. When the supply of nutrients is depleted, plasmodium either switches to a fructification state (if the level of illumination is high enough), when sporangia are produced, or forms a sclerotium (encapsulates itself in hard membrane), if in darkness. Therefore, we propose that a Physarum machine halts when all data nodes are utilized.

We can 'freeze' the computation by depriving plasmodium of water. In low-humidity conditions the plasmodium stops its foraging behavior and forms a sclerotium, a compact mass of hardened protoplasm. Results of the computation are not destroyed and remain detectable as 'empty/dead' protoplasmic tubes. In the state of sclerotium, the Physarum machine is ready for further deployment.

Formation of sclerotium can be seen as freezing of the storage structures, no nodes added or deleted, edges remain stationary. Fructification-type halting is equivalent to self-reproduction of a Physarum machine.

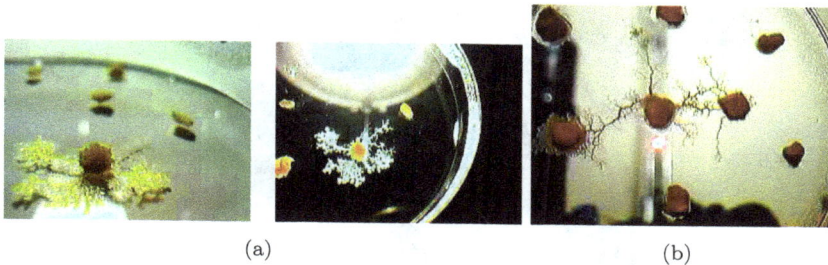

(a) (b)

Fig. 8.4 Active nodes and their developing active zones: (a) single active node is generating an active zone at the beginning of computation, (b) active zone around a central active node consisting of two nodes.

8.1.5 *Active zone*

In a KUM a storage graph must have an active node. This is an in-built feature of a Physarum machine. When plasmodium is grown in conditions with poor or no nutrients, then just one or a few nodes generate protoplasmic waves. In this case, the protoplasm spreads as mobile localizations, or localized wave fragments, by analogy with wave fragments in a subexcitable Belousov–Zhabotinsky medium [Sedina-Nadal et al. (2001)]. An example of a single active node, which has just started to develop its active zone, is shown in Fig. 8.4.

At every step of computation in a KUM there is an active node and some active zone, usually nodes neighboring the active node. The active zone has a limited complexity, in a sense that all elements of the zone are connected by some chain of edges to the initial node.

In general, the size of the active zone may vary depending on the computational task. In a Physarum machine, an active node is a trigger of contraction/excitation waves, which spread all over the plasmodium tree and cause pseudopodia to propagate, the shape to change and even protoplasmic veins to annihilate. The active zone comprises stationary and/or dynamic nodes connected to an active node with tubes of protoplasm.

A sample architecture of a Physarum machine is illustrated in Fig. 8.5. Some schematic representations of active zones are provided in Fig. 8.6.

8.1.6 *Bounded connectivity*

In contrast to Schönhage machines, KUMs have bounded in- and out-degrees of the storage graph. Graphs developed by Physarum are pre-

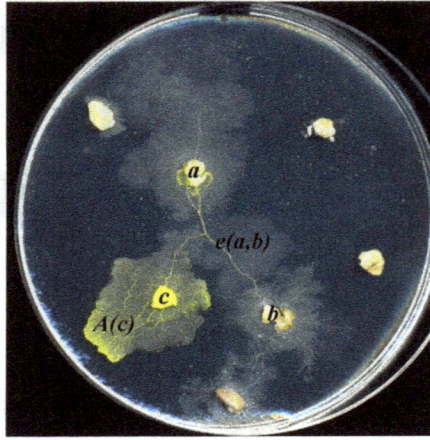

Fig. 8.5   An exemplary snapshot of a Physarum machine. Protoplasmic tube connecting flakes a and b represents an edge $e(a,b)$ of KUM. Active zone $A(c)$ emerges in the node c.

dominantly planar graphs. Moreover, if we put a piece of tube protoplasm on top of another tube of protoplasm, the veins fuse [Shirakawa (2007)]. Usually, not more than three protoplasmic strands join each other in one given point of space. It is reported that the average degree of the minimal spanning tree is around 1.99, and that of the relative neighborhood graph is around 2.6 [Cartigny et al. (2005)]. Graphs produced by standard procedures for generating combinatorial random planar graphs show a limited growth of average degree with the number of nodes or edges; the degree stays at around 4 when the number of edges increases from 100 to 4000 [Alber et al. (2001)]. We could assume that the average degree of a storage graph in Physarum machines is a little higher than the degree of spanning trees but less than the degree of random planar graphs.

8.1.7   *Addressing and labeling*

Every node of a KUM must be uniquely addressable and nodes and edges labeled: "the vertices joined by edges to a fixed vertex belong to different types" [Kolmogorov and Uspensky (1958)]. There are no direct implementations of such addressing in Physarum machines. With stationary nodes, this can be done either by coloring the nodes, or by tuning humidity of the oat flakes, thus changing the concentration of bacteria living on the flakes.

Examples of an experimental implementation of addressing and label-

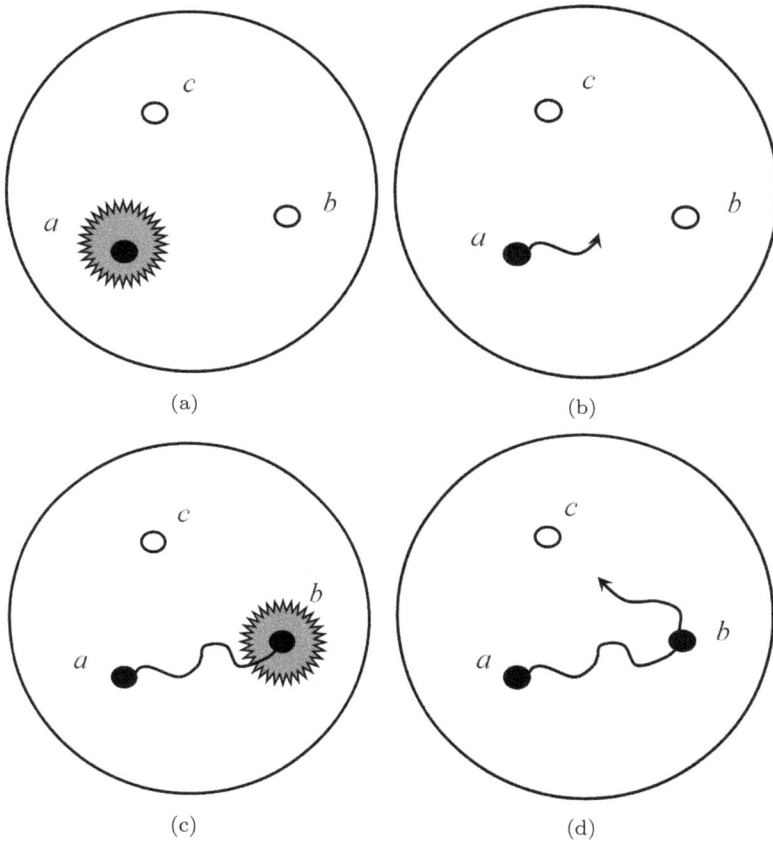

Fig. 8.6 Schematic dynamics of active zones: (a) active zone occupies node *a*, (b) active zone propagates towards node *b*, (c) active zone occupies node *b*, (d) active zone propagates towards node *c*. Oat flakes populated by plasmodium are shown by solid disks.

ing are shown in Fig. 8.7. Addressing a green-colored flake is shown in Fig. 8.7a, and addressing a blue flake with green-colored edge in Fig. 8.7b. Plasmodium does not treat all colors the same. So far, we have observed the following priority of choosing oat flakes, in descending order: uncolored, green, yellow/blue and red colored. This color preference allows plasmodium to selectively connect labeled nodes in the storage graph; an example is demonstrated in Fig. 8.7c.

(a)   (b)

(c)

Fig. 8.7   Addressing and labeling in a Physarum machine: (a) the machine is addressing a green-colored data node, (b) the edge out-coming from one data node is labeled in red, (c) plasmodium selectively connects green and yellow nodes in a storage structure.

8.1.8   *Basic operations*

A possible set of instructions for a Physarum machine could be as follows. Common instructions would include INPUT, OUTPUT, GO, HALT, and internal instructions would include NEW, SET, IF [Dexter et al. (1997)]. At the present state of experimental implementation, we assume that INPUT is done via distribution of sources of nutrients, while OUTPUT is recorded optically. Halting was discussed in previous sections. The SET instruction

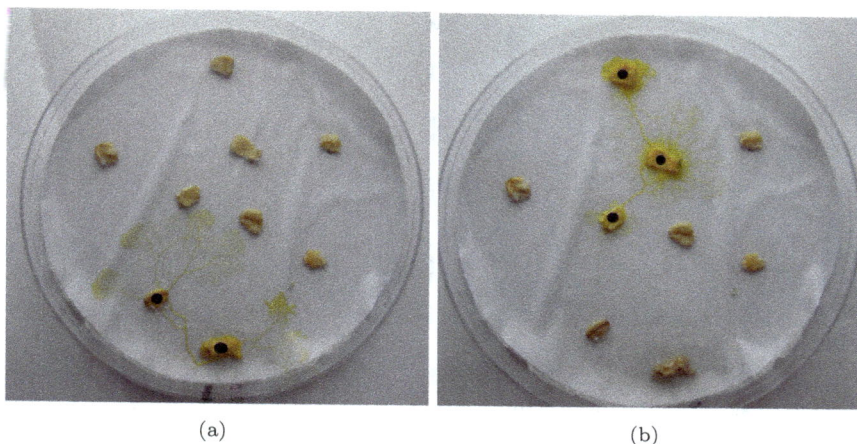

(a)   (b)

Fig. 8.8   Implementation of the REMOVE NODE operation by a Physarum machine. At first, plasmodium occupies two stationary nodes (a) at the southern part of the Petri dish, then removes all of these nodes from its storage structure by leaving them and occupying three stationary nodes at the northern part of the Petri dish (b). Nodes included in the storage structure are marked by solid disks.

causes pointers to redirect, and can be realized by placing a fresh nutrient source in the experimental container. When a new node is created, all pointers from the old node point to the new node.

Let us look at the experimental implementation of core instructions.

ADD NODE

To add a stationary node b to a node a's neighborhood, plasmodium must propagate from a to b, and form a protoplasmic tube representing the edge (ab). To form a dynamic node, propagating pseudopodia must branch into two or more pseudopodia, and the site of branching will represent a newly formed node; see example in Fig. 8.2.

REMOVE NODE

To remove a stationary node from a Physarum machine, plasmodium leaves the node (Fig. 8.8). To annihilate protoplasmic strands forming a dynamic node at their intersection, plasmodium removes the dynamic node.

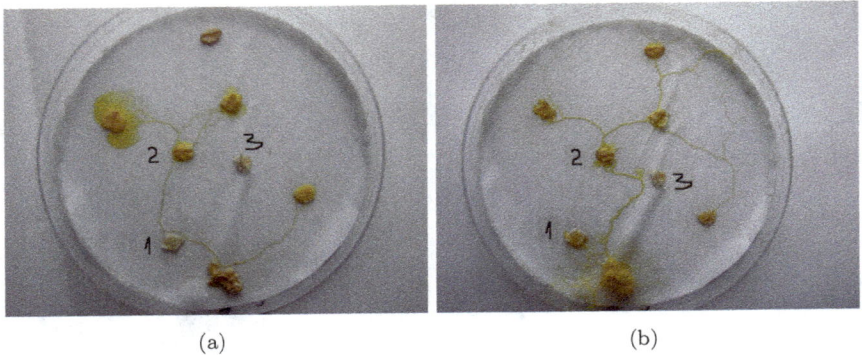

Fig. 8.9   Reconfiguration of the storage structure by a Physarum machine: implementation of ADD NODE, ADD EDGE and REMOVE EDGE operations.

ADD EDGE

To add an edge to a neighborhood, an active node generates propagating processes which establish a protoplasmic tube with one or more neighboring nodes. As an illustration, see formation of protoplasmic tubes between nodes marked by solid disks in Fig. 8.8.

REMOVE EDGE

When a protoplasmic tube annihilates, e.g. depending on global state or when a source of nutrients is exhausted, an edge represented by the tube is removed from the Physarum machine (Fig. 8.9). The following sequence of operations is demonstrated in Fig. 8.9: node 3 is added to the structure by removing an edge (12) and forming two new edges (13) and (23).

8.2   Example of Physarum machine solving simple task

Let us consider an example of a task solvable by a Physarum machine. Given nodes labeled red (R), green (G), blue (B) and yellow (Y), we want the plasmodium to establish protoplasmic tubes connecting the colored nodes in a chain.

A Physarum machine solves the task by first exploring the whole data space and then connecting required nodes (Fig. 8.10). The plasmodium tries to minimize its protoplasmic network. Sometimes auxiliary edges are formed, e.g. edges (OR) and (OG). They are removed later. The program,

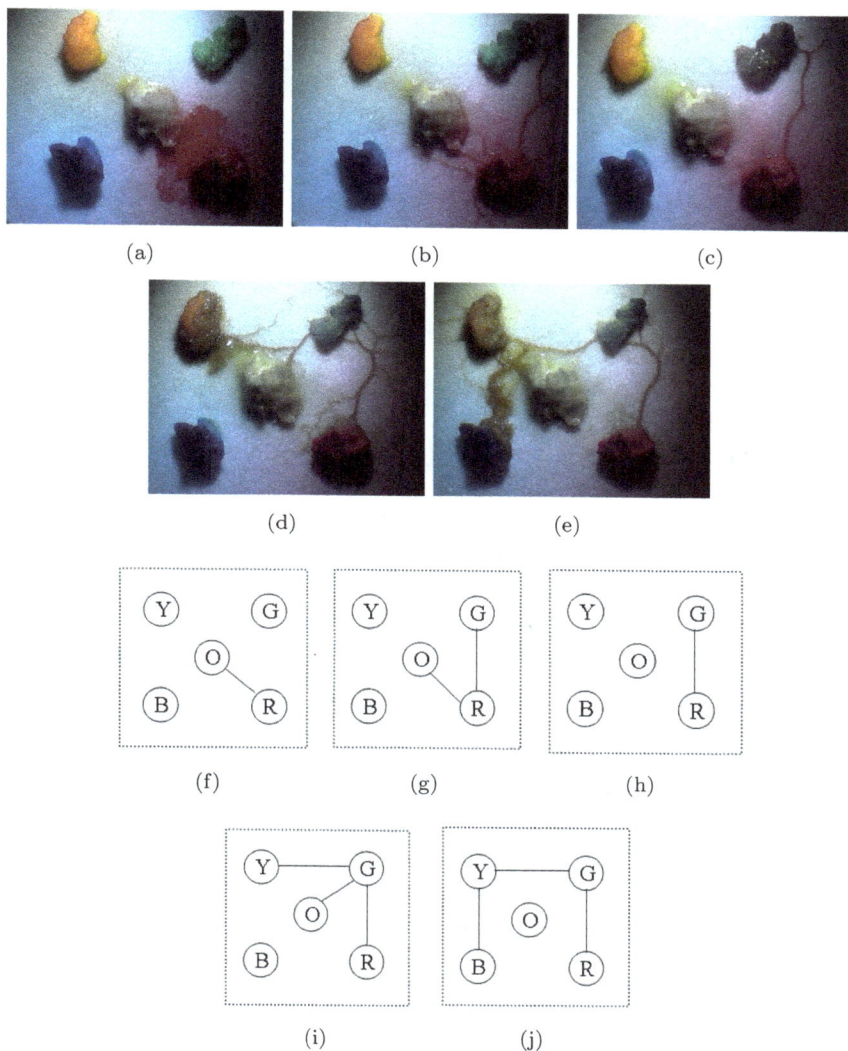

Fig. 8.10 Implementation of a simple task of connecting colored nodes by a Physarum machine: (a)–(e) show a sequence of photographs of plasmodium (magnification ×10) placed in a small container in the center of a virtual rectangle, whose corners are represented by oat flakes that are colored in yellow (Y), green (G), red (R), and blue (B); the scheme of the computation is shown in (f)–(j).

corresponding to Fig. 8.10, can be written as follows:

ADD EDGE(OR),
ADD EDGE(RG),
REMOVE EDGE(OR),
ADD EDGE(OG),
ADD EDGE(GY),
REMOVE EDGE(OG),
ADD EDGE(YB).

8.3   On parallelism

The KUM, in its original form, has a single control device; at least, 'modern age' interpretations [Gurevich (1988); Blass and Gurevich (2003)] claim that this is so. Does this apply to Physarum machines? Partly yes. Plasmodium acts as a unit in the long term, i.e. it can change position, or retract some processes in one place to form new ones in another place. However, periodic contractions of the protoplasm are usually initiated from a single source of contraction waves [Nakagaki et al. (2000); Tero et al. (2005)]. The source of the waves can be interpreted as a single control unit. However, sometimes during foraging behavior, several branches or processes of plasmodium can act independently and almost autonomously.

When thinking about parallelism in a Physarum machine, we can recall Bardzin's growing automata, outlined in 1964 [Bardzin (1964)], and developed in detail by Bardzin and Kalnins [Bardzin and Gurevich (2003)] (these are also utilized in causal networks [Gacs and Levin (1980)]). Bardzin's growing automata (BGA) are — in reality — parallel extensions of a KUM. In these automata, local transformations are simultaneously implemented in all nodes (and primitive operations are similar to that in a KUM).

A possible compromise between the original theoretical framework of a KUM and the partly parallel execution in experiments could be reached by proposing two levels of 'biological' commands executed by a Physarum machine's elements. There would have to be high-level commands, e.g.

- SEARCH FOR NUTRIENTS,
- ESCAPE FROM LIGHT,
- FORM SCLEROTIUM,
- FRUCTIFY,

and low-level commands, e.g.

- FORM PROCESS,
- PROPAGATE IN DIRECTION OF,
- OCCUPY SOURCE OF NUTRIENTS,
- RETRACT PROCESS,
- BRANCH.

Global commands are executed by plasmodium as a whole at once, i.e. in a given time step plasmodium executes only one high-level command. Local commands are executed by local parts of the plasmodium. Two spatially distant sites can execute different low-level commands at the same time.

8.4 Summary

We have suggested a possible biological system for realization of a Kolmogorov–Uspensky machine (KUM). In the experimental implementation we exploited the fact that in contrast to Turing machines (TMs), which read and react to symbols written on the tape, a KUM acts depending on the state of the neighborhood and its active node, including all neighboring nodes and edges.

Despite the overwhelming popularity of TMs, modern computing devices evolved from KUMs to Schönhage storage modification machines and other pointer machines, and then to random access machines. Possibly this happened because Kolmogorov and Uspensky clearly shifted focus from *algorithm* to *machine*:

> "... the schema for computing the value of a partially recursive function may not be directly given in the form of algorithm. If one develops this computation in the form of an algorithmic process ... then, through this, one automatically obtains a certain algorithm ..." [Kolmogorov and Uspensky (1958)].

We can speculate therefore that Physarum machines are the best biological realizations of current general-purposed computers.

Chapter 9

Reconfiguring Physarum machines with attractants

A program for a Physarum machine specifies how to replace the neighborhood of an active node (i.e. occupied by an active zone) with a new neighborhood, depending on the edges connected to the active node and the nodes in proximity to the active node [Blass and Gurevich (2003)]. In a Physarum machine the computation is implemented by the active zone, or several active zones. To make the computation process programmable, we must find ways of sensible and purposeful manipulation of the active zones. Several operations, i.e. manipulation procedures, are discussed in the present chapter.

9.1 Fusion and multiplication of active zones

Schematic examples of fusion and multiplication operations are shown in Fig. 9.1. When two active zones A_1 and A_2 fuse they can produce either a new active zone A_3 (Fig. 9.1a) or just a protoplasmic tube (Fig. 9.1b). Multiplication of active zones is another basic operation: the active zone A_1 splits into two independent active zones A_2 and A_3 propagating along their own trajectories (Fig. 9.1c). In some cases active zones can be generated from 'empty active zones', oat flakes populated by plasmodium (Fig. 9.1d).

Finding 11.   *The operation* FUSE(A_1, A_2), *fusing of two active zones A_1 and A_2, can be implemented via collision of the active zones attracted by sources of nutrients.*

An example is demonstrated in Fig. 9.2.
Five oat flakes were arranged in a line on wet filter paper. Two pieces of plasmodium were placed near the extreme flakes, one piece near the south-

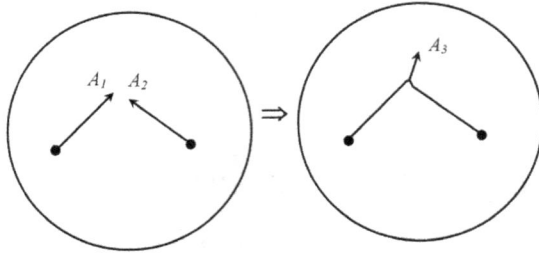

(a)

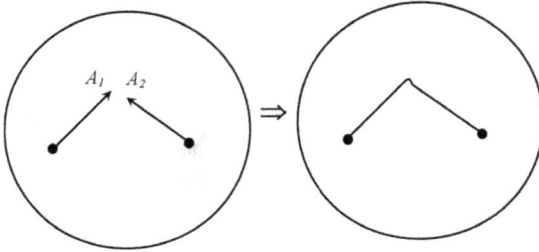

(b)

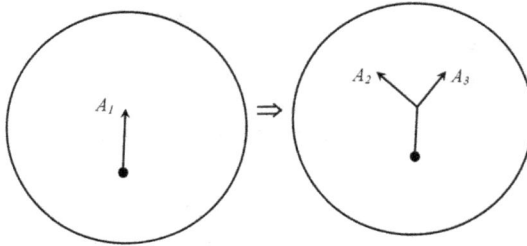

(c)

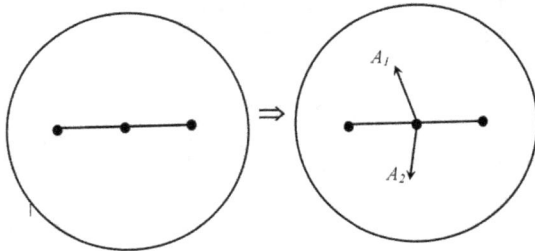

(d)

Fig. 9.1 Schematics of operation with active zones: (a) FUSE(A_1, A_2) = ∅, (b) FUSE(A_1, A_2) = A_3, (c) MULT(A_1) = $\{A_2, A_3\}$, (d) MULT(∅) = $\{A_1, A_2\}$.

(a) t=0 h

(b) t=12 h

(c) t=24 h

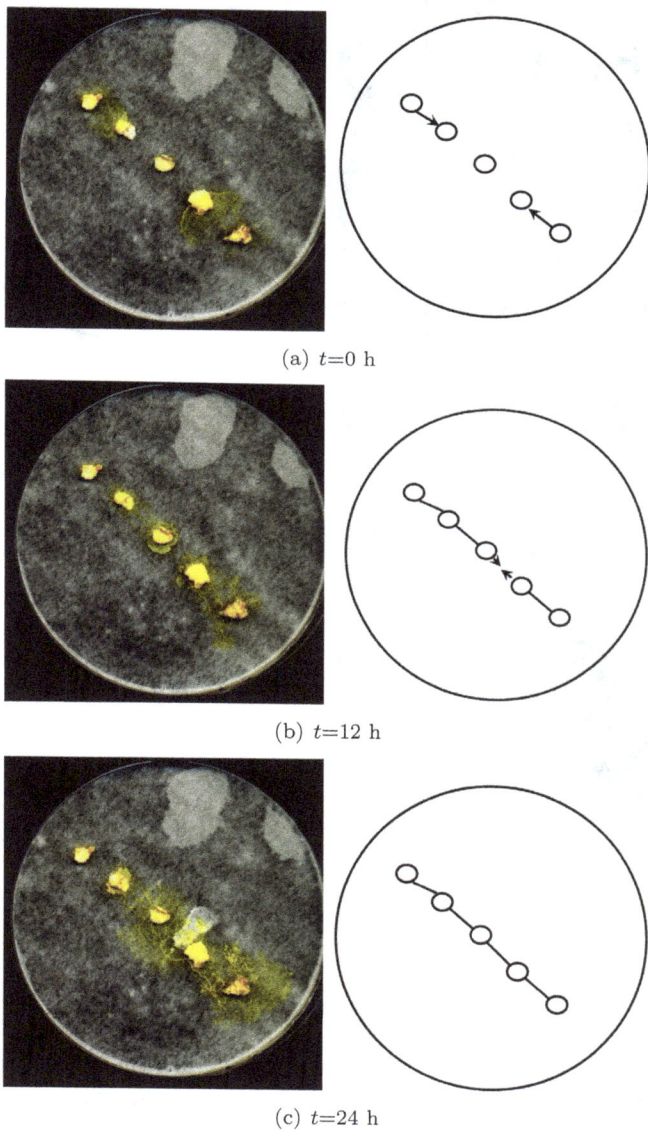

Fig. 9.2   Fusion of active zones. Protoplasmic tubes are shown as lines and active zones as arrows in the accompanying diagrams.

(a) $t = 0$ h

(b) $t = 10$ h

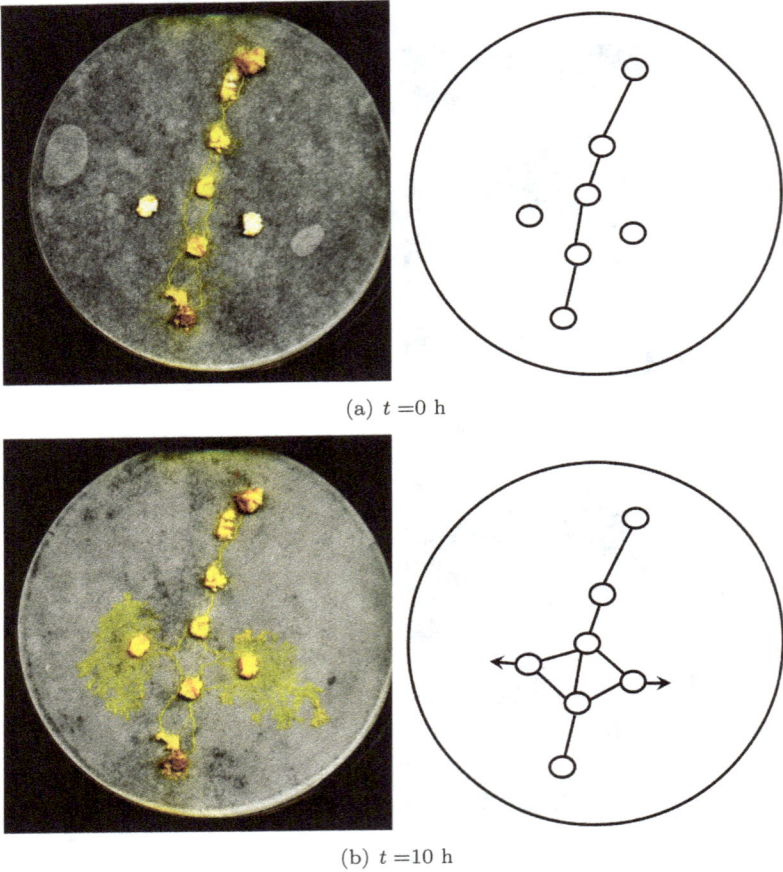

Fig. 9.3   Multiplication of active zone. Protoplasmic tubes are shown as lines and active zones as arrows in the accompanying diagrams.

western flake, the other piece near the north-eastern flake. The active zones are formed as pseudopodia and propagate towards the center of the chain (Fig. 9.2a and b). When the active zones A_1 and A_2 collide they fuse and annihilate, FUSE$(A_1, A_2) = \emptyset$ (Fig. 9.2c). Depending on the particular circumstances, the new active zone (the result of fusing) may become inactive, transform to protoplasmic tubes or remain active, FUSE$(A_1, A_2) = A$, and continue propagation in a new direction.

Finding 12.   *Multiplication of an active zone, the operation* MULT$(A) = \{A_1, A_2\}$, *can be implemented by placing sources of nutrients near the pro-*

to⸱lasmic tubes, or inactive zones.

An example is shown in Fig. 9.3. A chain of oat flakes is connected by protoplasmic tubes; we add two new oat flakes to evoke new active zones (Fig. 9.3a). Ten hours later two active zones A_1 and A_2 are formed; each pseudopodium travels to its unique oat flake (Fig. 9.3b).

9.2 Translating active zone

In some cases we may need to translate the active zone not to another graph node (source of nutrients) but to a domain of an 'empty' space. Moreover, it may be necessary to provide the active zone A with a certain initial velocity vector $\mathbf{v}$ so that the zone continues its propagation in the pre-determined direction. Such an operation DIRECT$(A, \mathbf{v})$ is executed using additional sources of nutrients (Fig. 9.4). Given a chain of oat flakes connected by protoplasmic tubes (Fig. 9.4a), we added two new flakes to the west of the chain. A pseudopodium, representing the active zone A, sprouted from the 'old' flake lying between projections of the two new flakes. It continues propagating along the bisector separating the two new flakes. The active zone continues its propagation towards the west until it collides with the wall of the Petri dish. Meanwhile, two more active zones are formed to connect the new oat flakes to the existing protoplasmic network (Fig. 9.4b).

Finding 13. *Translation of an active zone may lead to, or be used for, major restructuring of the data storage graph.*

Thus, a parallel partial shift of the graph chain is demonstrated in Fig. 9.5. Three new oat flakes are placed alongside the chain of oat flakes already connected by protoplasmic tubes (Fig. 9.5a). In a few hours the plasmodium occupies new oat flakes. Moreover, part of the protoplasmic graph co-aligned with new oat flakes is shifted to new oat flakes (Fig. 9.5b). Abandoned protoplasmic tubes, former edges of the shifted part, are visible as white tubes.

Such a major restructuring of the graph was caused by relocation of the active zone. At first, the active zone traveled north-east (Fig. 9.5a). Addition of new oat flakes caused the zone to switch to the new location and move south-west (Fig. 9.5b).

(a) $t = 0$ h

(b) $t = 12$ h

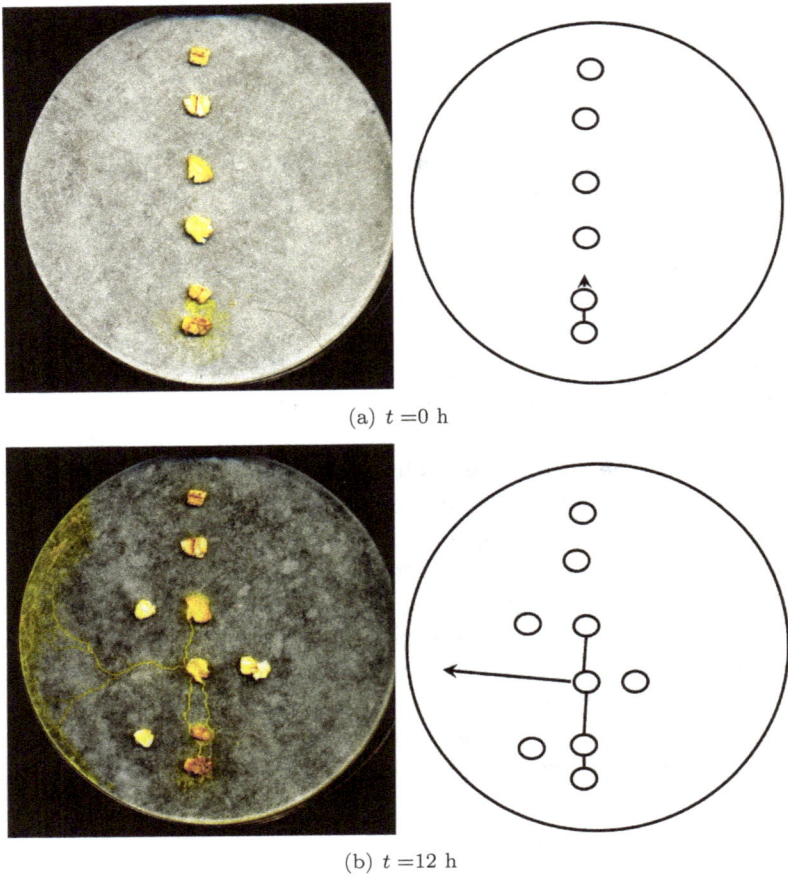

Fig. 9.4 Relocation of active zones to a domain without food sources, operation DIRECT$(A, \mathbf{v})$. Protoplasmic tubes are shown as lines and active zones as arrows in the accompanying diagrams. Vector $\mathbf{v}$ is shown by westward-pointing arrow in (b).

9.3 Reconfiguration of Physarum machine

In many examples above we observed the formation of protoplasmic tubes, connecting two geographically closest sources of nutrients. Given a chain of oat flakes, an additional oat flake is added to the experimental container. What is the exact mechanism of inclusion of the new flake in the graph? Do all nodes of the graph develop active zones, which travel to the flake while competing with each other for the flake?

(a) $t = 8$ h

(b) $t = 19$ h

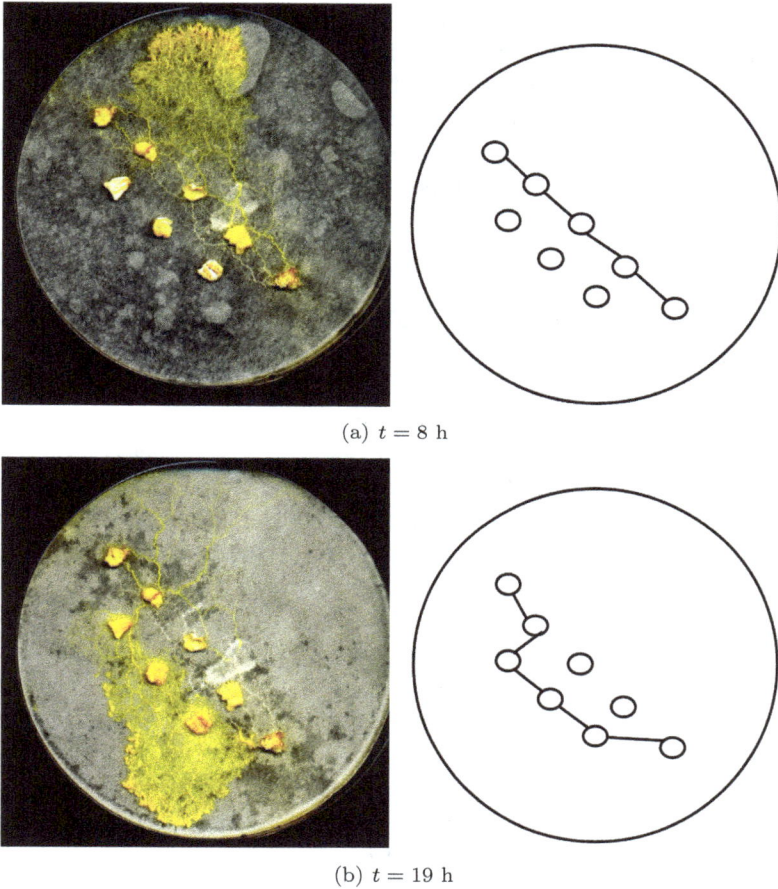

Fig. 9.5   Illustration of major restructuring of data storage graph.

Our experimental observations show the following.

Finding 14. *Let a Physarum machine span an experimental space with a graph G of its protoplasmic network and let one new source of nutrients be added to the Physarum's environment. If the graph G is connected, then only one active zone, heading for the new source of nutrients, emerges. If the graph G has two or more connected components, then two or more active zones are generated.*

This happens due to sychronization of activity in the whole protoplasmic network.

This may indicate that the plasmodium first 'decides' which part of the protoplasmic graph is closest to the recently added source of nutrients and only then generates an active zone. All parts of the Physarum machine sense the chemo-attractants coming from the new source of nutrients; however, the node closest to the new source somehow inhibits activity of other nodes. Existence of connections between nodes of the Physarum machine is a prerequisite for the inhibition.

To prove the point, we arranged oat flakes in a chain and allowed the chain to be spanned by the plasmodium (Fig. 9.6a). When the flakes are connected by protoplasmic tubes we added a new oat flake, the eastern-most flake in the picture (Fig. 9.6a).

To break communication between nodes of the Physarum machine, we cut through protoplasmic tubes connecting three northern nodes, (cuts are shown by lines in Fig. 9.6a). Due to breakup in communication all nodes of the plasmodium react, almost simultaneously, to the addition of a new source of nutrients by sprouting pseudopodia (Fig. 9.6b). In a few hours the cut tubes are restored (Fig. 9.6c). This reinstates communication between all parts of the Physarum machine.

The distant, from the new flake, nodes of the Physarum machine deactivate their active zones and cease propagation of pseudopodia. As the result of restored communication only one active zone remains, and the new node is connected to the storage graph of the Physarum machine by a single edge.

Just cutting protoplasmic tubes does not lead to formation of new pseudopodia (Fig. 9.7). When tubes are cut they do usually fuse back again in a few hours without any adverse effect on the plasmodium behavior.

9.4 Summary

We outlined basic procedures for control of active zones of a Physarum machine, an implementation of a general-purpose Kolmogorov–Uspensky machine (KUM). We were only concerned with manipulating active zones (actively growing pseudopodia of the plasmodium) because they are the main computational units of the storage modification machine, implemented in the plasmodium. In laboratory experiments we executed basic operations with active zones. We showed how to merge two active zones, to multiply an active zone, to translate an active zone to a new node of the storage structure and to direct an active zone in the space not occupied by data

(a) $t=0$ h

(b) $t=11$ h

(c) $t=17$ h

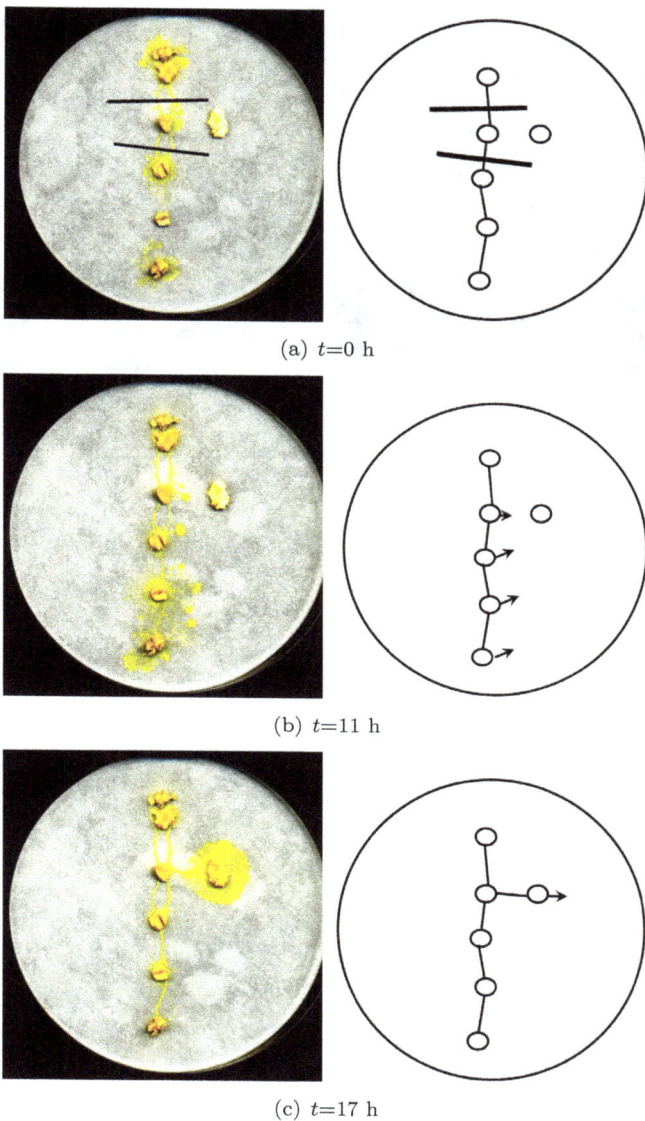

Fig. 9.6 Cutting an edge leads to desynchronization of the Physarum machine. Protoplasmic tubes are shown as lines and active zones as arrows in the accompanying diagrams.

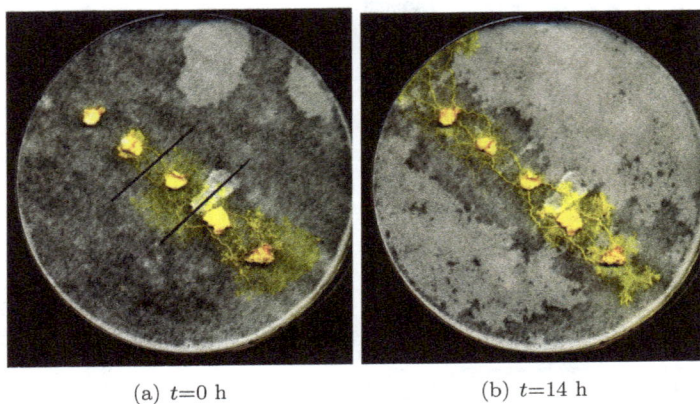

(a) t=0 h   (b) t=14 h

Fig. 9.7   Absence of adverse reaction to tube cutting.

nodes. We envisage that our experimental findings will be employed in
future programming of spatially distributed biological computers.

Chapter 10

Programming Physarum machines with light

A key component of the Kolmogorov–Uspensky machine (KUM) is an active zone [Kolmogorov (1953); Uspensky (1992)], which may be seen as a computational equivalent to the head in a Turing machine. Physical control of the active zone is of utmost importance because it determines functionality of the biological storage modification machine. In Chap. 8.4, we have shown how to manipulate active zones using positive stimuli, by positioning sources of nutrients. Now we consider negative stimulation. We experimentally demonstrate that propagation of an active zone can be tuned by localized domains of illumination.

10.1 Physarum and light

The plasmodium of *P. polycephalum* exhibits negative photo-taxis. The general understanding of the plasmodium's response to light is that the plasmodium moves away from light when it can and switches to another phase of its life cycle or undergoes fragmentation when it could not escape from light. If a plasmodium, particularly a starving one [Guttes (1961)], is subjected to a high intensity of light the plasmodium turns into a sporulation phase [Sauer et al. (1969)]. There is evidence that phytochromes are involved in the light-induced sporulation [Starostzik and Marwan (1995)] and a sporulation morphogene is transferred by protoplasmic streams to all parts of the plasmodium [Hildebrandt (1986)].

Photo-fragmentation is another physiological response to strong and unavoidable illumination. When a plasmodium is illuminated by ultraviolet or blue monochromatic light, in a hostile environment of laboratory conditions, it breaks up into many equally sized fragments (each fragment contains around eight nuclei) [Kakiuchi et al. (2001)]. The fragmentation is transient and after some time the fragments merge back into a fully functional plasmodium.

In this chapter we focus on photo-movement, a less drastic response to illumination than the two scenarios mentioned above. Plasmodium of *P. polycephalum* exhibits the most pronounced negative photo-taxis to blue and white light [Bialczyk (1979); Schreckenbach et al. (1984)]. The illumination increase causes changes in the plasmodium's oscillatory activity; the degree of changes is proportional to the distance from the light source [Wohlfarth-Bottermann and Block (1981); Block and Wohlfarth-Bottermann (1981)]. The exact mechanism of the response to light is as yet unknown. There are however a few phenomena uncovered in experiments. The first is the presence of phytochrome-like pigments [Kakiuchi et al. (2001)], which might be primary receptors of illumination. The light response of the pigments triggers a chain of biochemical processes [Schreckenbach et al. (1981)]. These processes include increase in activity of isomerase enzymes [Starona and Wojciech (1992)], changes in mitochondrial respiration [Korohoda et al. (1983)] and spatially distributed oscillations in ATP concentrations [Ueda et al. (1986)].

Nakagaki et al. [Nakagaki et al (1999); Nakagaki et al. (2007)] undertook the first ever experiments on shaping plasmodium behavior with illumination. They discovered that protoplasm streaming oscillations of plasmodium can be tuned by, or relatively synchronized with, periodic illumination [Nakagaki et al (1999)]. They also demonstrated that plasmodium optimizes its protoplasmic network structure in a field with heterogeneous illumination [Nakagaki et al. (2007)]: the thicknesses of protoplasmic tubes in illuminated areas are less than the thicknesses of tubes in shaded areas [Nakagaki et al. (2007)]. These indicate that illuminated domains could make a convenient tool to input instructions to different parts of Physarum machines in parallel.

Several basic questions need to be answered. What exactly does a pseudopodium or a migrating Physarum do when they approach an illuminated domain? How can light be used to program plasmodium movements? What types of plasmodium reflections can be implemented using light mirrors? Can we shape a structure of a plasmodium network by heterogeneous illumination? In this chapter we present experimental and theoretical findings we obtained while trying to answer these questions.

Petri dish with 2 mm agar gel and plasmodium

Opaque plate with transparent shape cut in

Electro-luminescent sheet

(a)

(b)

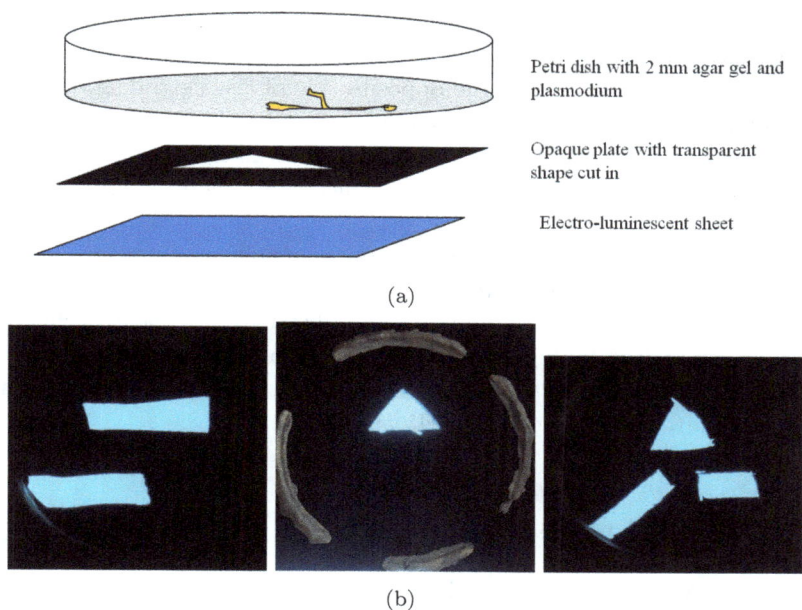

Fig. 10.1 Experimental setup: (a) scheme, (b) photographs of illuminating domains, i e. geometrical shapes.

10.2 Designing control domains

Experiments on controlling the plasmodium with light were undertaken in standard Petri dishes, 9 cm in diameter. A substrate was 2% agar gel. Light obstacles, or illumination domains, were implemented using electro-luminescent sheets[1]. Based on previous works on negative photo-taxis of the plasmodium, we have chosen blue illuminating sheets. The nominal sheet's brightness (at 110 V, 400 Hz AC) is 73 cd/m^2 (blue). The electro-luminescent sheets do not produce heat; therefore, they can be positioned in close contact with the growing substrate. We prepared the masks by cutting rectangles and triangles in black plastic (bottoms of CD pack spindles and black packaging of microwaveable food). When a mask is placed on top of the electro-luminescent sheet, the light is passed only through the cuts (Fig. 10.1).

The experiments were conducted in a room with diffusive light of 3–

[1]Manufacturer Seikosha, supplier RS Components Ltd, Birchington Road, Corby, Northants, NN17 9RS, UK.

5 cd/m, 22°C temperature. In each experiment an oat flake colonized by the plasmodium was placed on one side of a Petri dish, and a few oat flakes without plasmodium at the opposite side of the Petri dish (to shape directional propagation of the plasmodium). Petri dishes with plasmodium were periodically scanned on a standard HP scanner. The only editing done to the scanned images is color enhancement: increase of saturation and contrast.

We simulate spatio-temporal dynamics of plasmodium on a non-nutrient substrate using the two-variable Oregonator equations introduced in Chap. 4.3. the illuminated domain (light obstacle) S of the experimental space was simulated by higher values of the parameter ϕ: if a given point belongs to S, then $\phi = 0.085$ (the medium inside the light obstacle becomes non-excitable), otherwise $\phi = 0.0804$.

10.3   Trees and waves

There are two distinct forms of the plasmodium: a protoplasmic tree and a traveling localization.

The protoplasmic tree is formed when plasmodium forages the space by sprouting pseudopodia in various directions. The pseudopodia remain connected to the original location, and still the main 'body', of the plasmodium by protoplasmic tubes.

In certain situations the plasmodium leaves its original location and propagates as a whole, i.e. migrates, on the substrate. A migrating plasmodium is shaped similarly to a wave fragment traveling in a sub-excitable medium [Adamatzky et al. (2008); Adamatzky (2009)]. The boundary between the two morphologies may be fuzzy, and very often we can observe wave fragment like shapes of pseudopodia, and gradual transitions between tree-like and wave fragment like morphology.

An example of the 'tree to wave fragment' transformation is shown in Fig. 10.2. An oat flake colonized by plasmodium is placed in the southern part of a Petri dish (Fig. 10.2a). The plasmodium sprouts several pseudopodia exploring the space (Fig. 10.2a). The protoplasmic branches die out when they encounter illuminated domains (Fig. 10.2b and c). A group of branching pseudopodia tries to find a way around the triangular light obstacle (Fig. 10.2c) but eventually abandons the attempt (Fig. 10.2d).

By that time, bacteria on the original oat flakes are exhausted and the plasmodium switches to its migration phase (Fig. 10.2d). The plasmodium

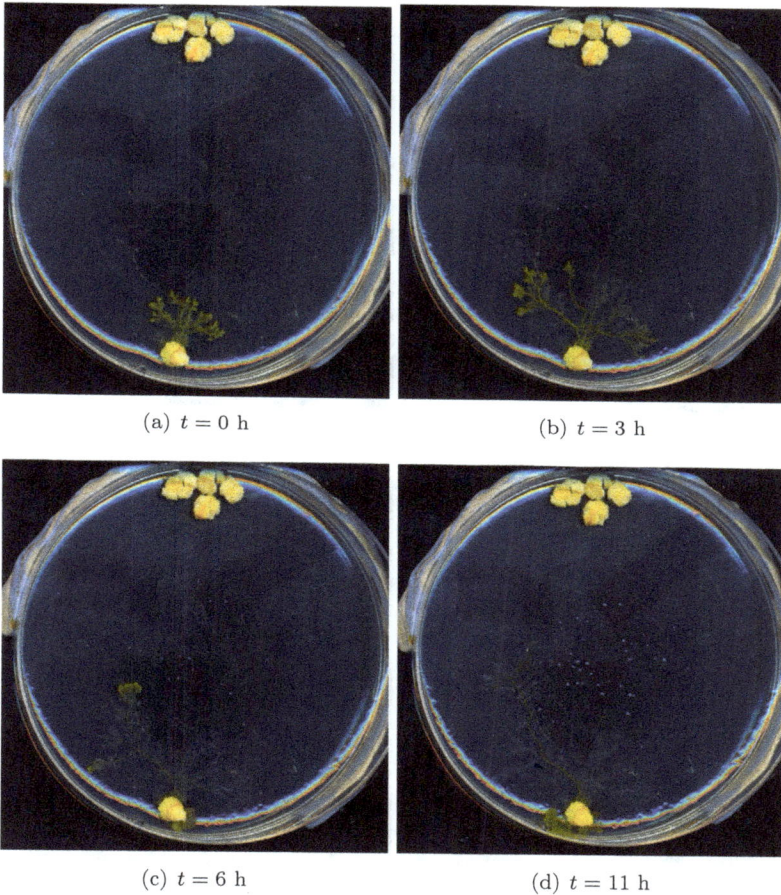

(a) $t = 0$ h

(b) $t = 3$ h

(c) $t = 6$ h

(d) $t = 11$ h

Fig. 10.2 Transformation of plasmodium from protoplasmic tree to traveling localization, i.e. migrating plasmodium. Scanned images of experimental Petri dish.

abandons its original oat flake and starts propagating *as* a whole along the eastern wall of the Petri dish (Fig. 10.2e and f). A typical wave fragment of the propagating plasmodium is formed (Fig. 10.2g), which heads towards the source of nutrients (a group of oat flakes in the northern part of the Petri dish). Eventually, the plasmodium reaches a new source of nutrients (Fig. 10.2h).

When the plasmodium propagates as *a* whole, it looks like and behaves like a wave fragment in a sub-excitable medium [Adamatzky et al. (2008);

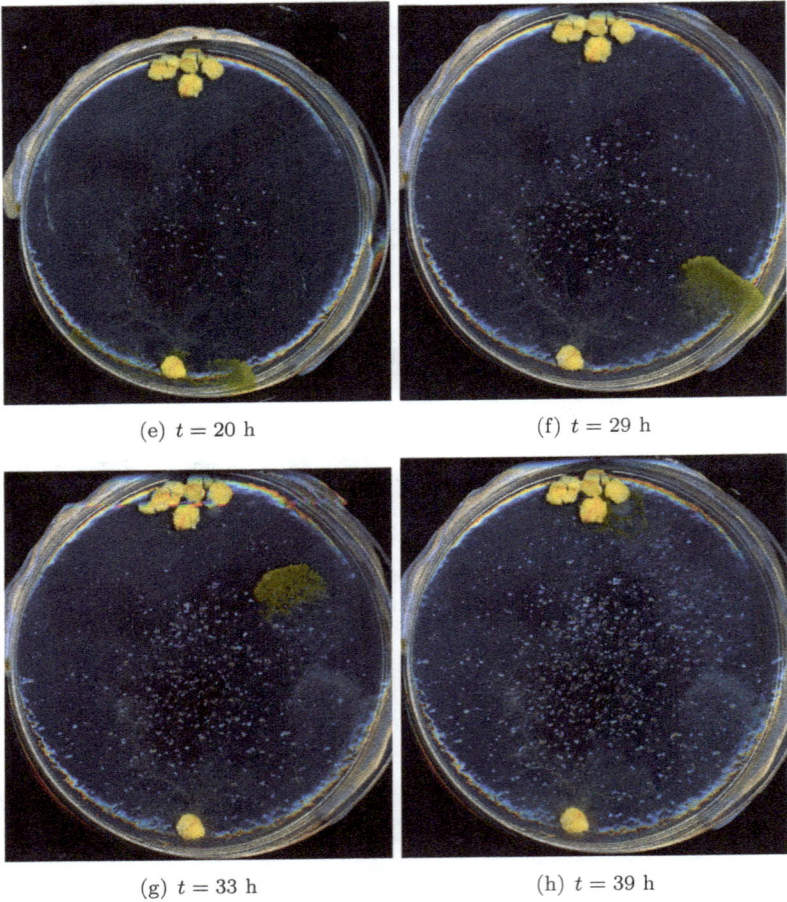

(e) $t = 20$ h   (f) $t = 29$ h

(g) $t = 33$ h   (h) $t = 39$ h

Fig. 10.2   *Continued.*

Adamatzky (2009)]. Moreover, a migrating plasmodium is more sensitive to potential environmental threats than just a propagating pseudopodia. This is because a protoplasm in the pseudopodia can always be retracted back to the main body of the plasmodium; thus, pseudopodia can 'take risks'. The migrating plasmodium is more vulnerable, because any 'miscalculation' in choosing its migration route may lead to disaster.

10.4   Diverting plasmodium

In an ideal situation the plasmodium propagates as a wave fragment and —
unless it encounters an obstacle on the way — keeps its shape and velocity
vector conserved. If a 'head' of the plasmodium wave comes upon a highly
illuminated domain, the frequency of protoplasm oscillations in this domain
increases. Due to a difference in the protoplasm oscillation frequency, the
plasmodium wave slightly turns to the side with less oscillating protoplasm.

Proposition 10.1. *We can steer propagating plasmodium using light obstacles.*

   A deflection of the plasmodium wave by a light triangle is demonstrated in Fig. 10.3. The plasmodium wave propagates north-north-east
(Fig. 10.3a and e). The plasmodium hits a light triangle on its western
side (Fig. 10.3b and f). The light increases the frequency of oscillations
in the illuminated part of the plasmodium. The plasmodium wave turns
eastwards (Fig. 10.3c and g) and travels in the new direction until it hits
the dish's wall (Fig. 10.3c, d, h).

10.5   Inertia

The plasmodium may not reflect from a light obstacle immediately. After
encountering even a highly illuminated area, the plasmodium wave carries
on traveling in the original direction and only after passing the obstacle
starts diverting. Such an example of delayed reaction to illumination is
shown in Fig. 10.4. A plasmodium wave, originated from the plasmodium
colonizing a group of flakes in the northern part of the dish, travels south
(Fig. 10.4a and c). The wave passes through the rectangular illuminated
area without any immediate reaction (Fig. 10.4a). The response occurs
several hours later. The plasmodium steers south-west (Fig. 10.4b and d).
Note that the plasmodium is *not* diverted by the second, lying southward,
light rectangle, because when the diversion happens the plasmodium is far
from the second light rectangle.

   Another example of the steering by light obstacles is shown in Fig. 10.5.
The plasmodium wave, heading north, enters the illuminated rectangular
area (Fig. 10.5a and c). The wave is slightly displaced towards the eastern
side of the light rectangle. Due to differences in light-induced oscillation
of protoplasm, the plasmodium turns north-east and continues traveling

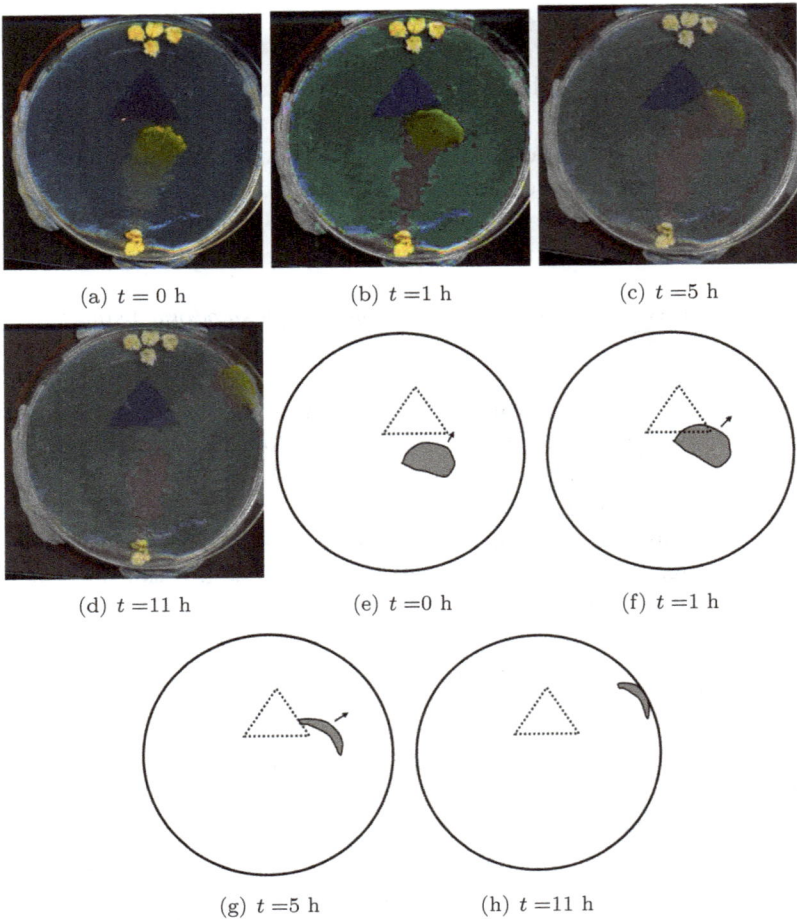

(a) $t = 0$ h (b) $t = 1$ h (c) $t = 5$ h

(d) $t = 11$ h (e) $t = 0$ h (f) $t = 1$ h

(g) $t = 5$ h (h) $t = 11$ h

Fig. 10.3 Deflection of a plasmodium wave by light triangle: (a)–(d) scanned images of the experimental Petri dish, illuminated triangular cut is seen as darker domain, (e)–(h) schemes of the propagating plasmodium waves.

in this direction until it collides with a wall of the Petri dish (Fig. 10.5b and d).

Finding 15. *By combining light obstacles and chemo-attractants, we can implement accurate control of plasmodium waves.*

The plasmodium is attracted to oat flakes populated with bacteria and the plasmodium senses and reacts to sources of nutrients placed as far as

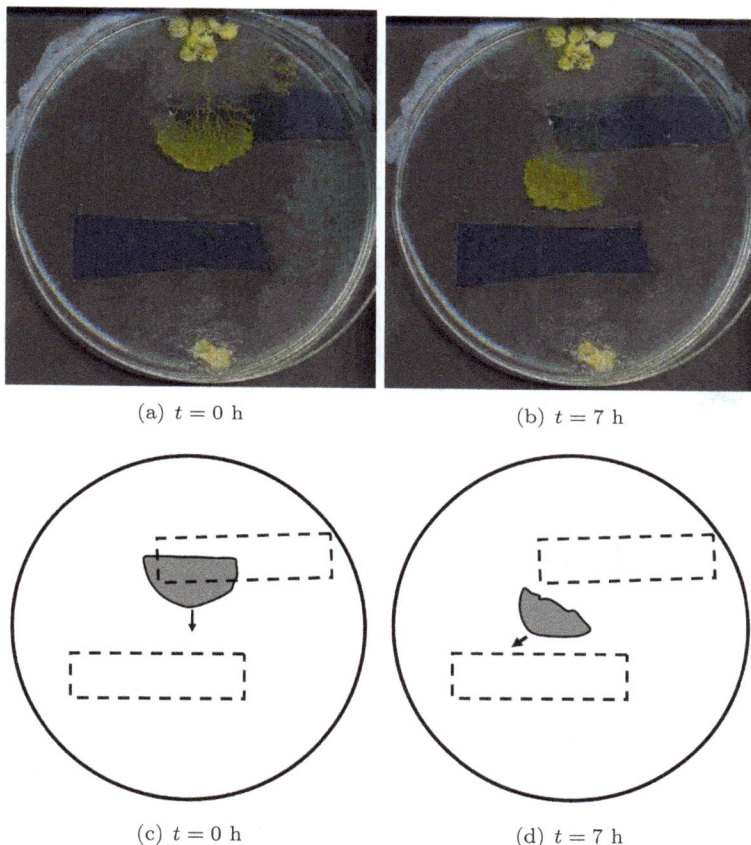

(a) $t = 0$ h (b) $t = 7$ h

(c) $t = 0$ h (d) $t = 7$ h

Fig. 10.4 Deflection of plasmodium wave by light rectangle: (a) and (b) photographs of the plasmodium, (c) and (d) schemes of the propagating plasmodium waves.

3–4 cm away. The plasmodium 'ascends' along gradients of the attractants until it reaches the sources of nutrients. By placing light obstacles in the attracting field, we can tune and shape the trajectory of the plasmodium's motion. For example, the plasmodium shown in Fig. 10.6 is reflected eastwards by the first illuminated rectangle it encounters (Fig. 10.6a and d). Later, the plasmodium turns north, being attracted to oat flakes in the northern part of the Petri dish. The plasmodium is directed westwards by the second illuminated rectangle (Fig. 10.6b and e). However, it does not continue west but turns north again to reach the sources of nutrients (Fig. 10.6c and f).

(a) $t = 0$ h   (b) $t = 12$ h

(c) $t = 0$ h   (d) $t = 12$ h

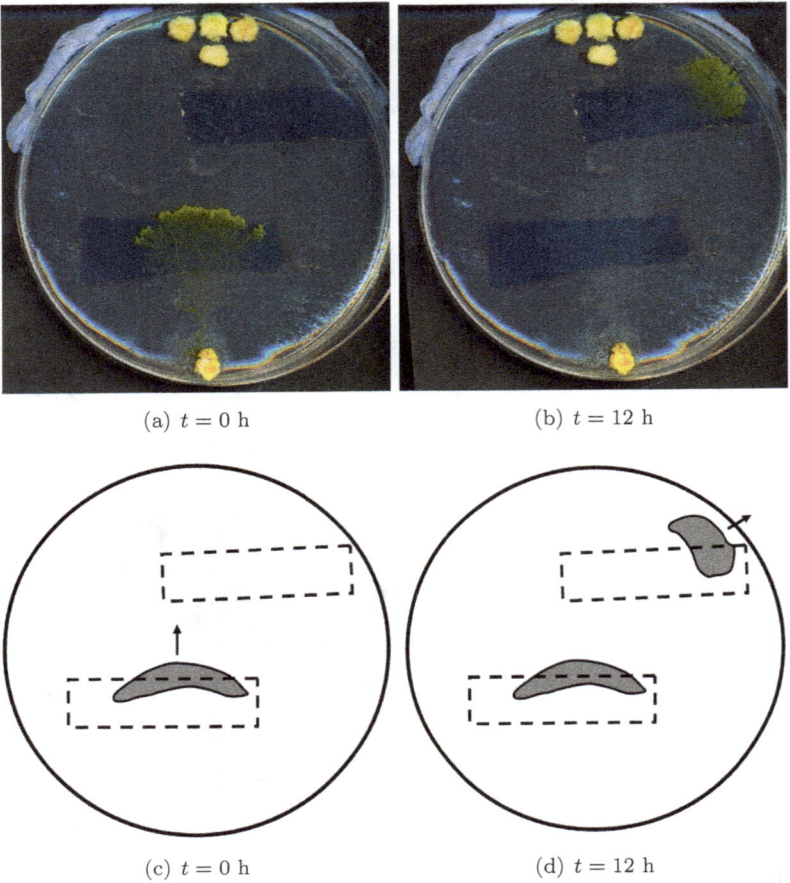

Fig. 10.5   Deflection of plasmodium wave by light rectangle: (a) and (b) photographs of the plasmodium, (c) and (d) schemes of the propagating plasmodium waves.

The Oregonator model of the plasmodium waves impeccably matches our experimental results. An example of simulation is shown in Fig. 10.7.

At the beginning of the experiments, a piece of plasmodium is simulated by exciting the medium with an ellipse-shaped stimulus (Fig. 10.7a). The medium is perturbed by an initial excitation, when 15×15 sites are assigned $u = 1.0$ each.

The perturbation generates an ellipse-shaped excitation wave. Due to sub-excitability of the simulated medium, the initially circular wave front

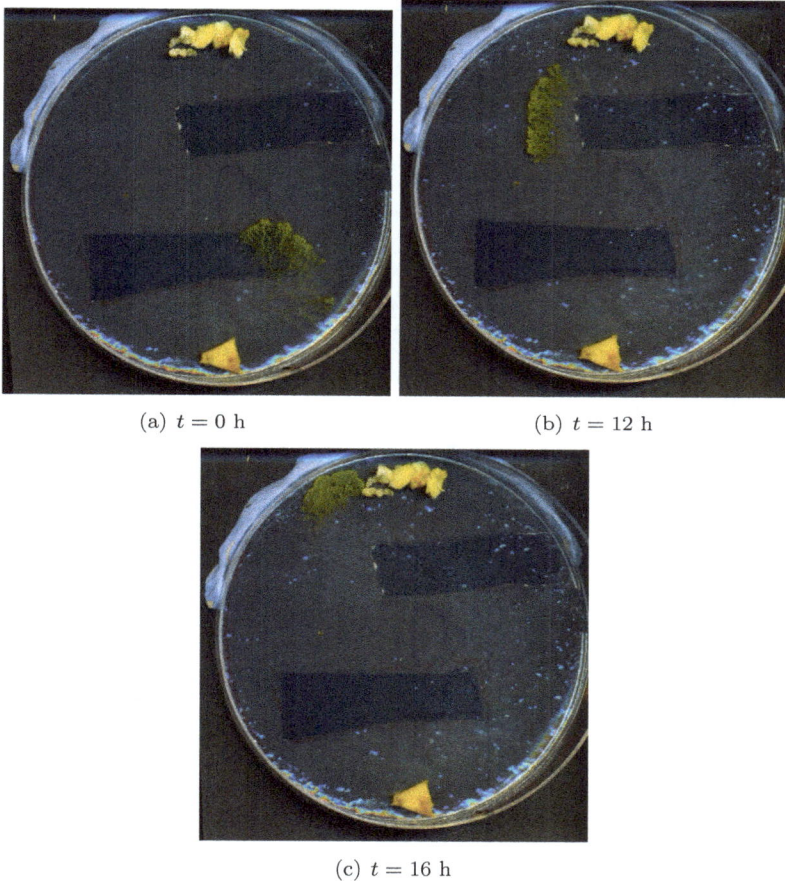

(a) $t = 0$ h   (b) $t = 12$ h

(c) $t = 16$ h

Fig. 10.6 Steering plasmodium with combination of light obstacles and attracting field: (a)–(c) scanned images of experimental Petri dish, (d)–(f) schemes of the propagating plasmodium.

(Fig. 10.7b) breaks up into two wave fragments. One wave travels north, the other wave south (Fig. 10.7c–e). We discard the south-traveling wave. When the eastern part of the north-traveling wave enters the illuminated area, it becomes inhibited and gradually disappears (Fig. 10.7e and f). The remaining part of the wave continues its travel as an individual wave, propagating north-north-west (Fig. 10.7g–h).

(d) $t = 0$ h   (e) $t = 12$ h

(f) $t = 16$ h

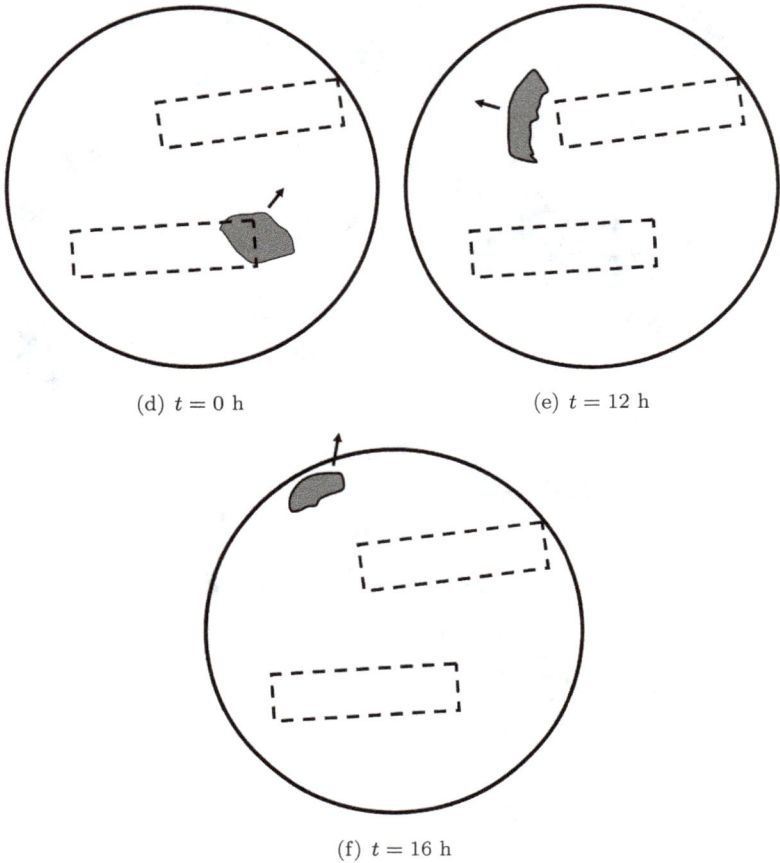

Fig. 10.6   *Continued.*

10.6   Multiplying plasmodium waves

Finding 16.   *A propagating plasmodium wave or a pseudopodium can be split by a suitably shaped domain of illumination.*

In some situations, propagating plasmodium hits a light obstacle which is small enough to divert the whole plasmodium wave. If parts of the plasmodium wave remain outside the illuminated shape, these parts continue to travel as 'independent' plasmodium waves. Thus, the plasmodium wave becomes split into two waves. See the example in Fig. 10.8. In this par-

(a) $t = 200$ (b) $t = 400$ (c) $t = 600$ (d) $t = 1000$

(e) $t = 1200$ (f) $t = 1400$ (g) $t = 1600$ (h) $t = 1800$

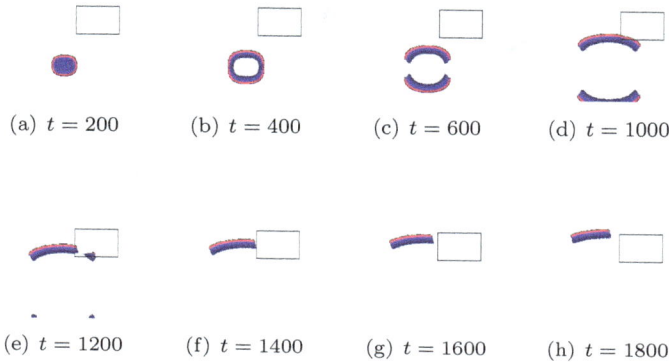

Fig. 10.7 Simulation of plasmodium wave diverted by rectangular illuminated domain. Red and blue components of each pixel's color are defined as follows. If $u > 0.1$, then red value $255 \cdot u$, if $v > 0.1$, then blue value $600 \cdot v$, otherwise the background white color is used. Time shows steps of numerical integration.

ticular experiment, the plasmodium wave comprises several pseudopodia moving together in a close group. On its way towards the source of nutrients, the wave runs across the illuminated triangle (Fig. 10.8a and d). The pseudopodia try to steer away from the source of light and move towards the western and eastern sides of the illuminated triangle (Fig. 10.8b and e). Eventually, two separate groups of pseudopodia are formed; one group travels north-west, the other group moves north-east (Fig. 10.8b and e).

Simulation results for wave splitting are shown in Fig. 10.9. An ellipse-shaped zone of initial excitation (Fig. 10.9a) produces two wave fragments traveling north and south (Fig. 10.9b and c). We are concerned with the fragment propagating north (Fig. 10.9de). When the wave fragment reaches the illuminated triangle the part of the wave inside the triangle extinguishes due to excitability inhibited by light (Fig. 10.9f and g). As a result of the inhibition, two separate wave fragments are formed. One wave fragment travels south-west, the other travels south-east (Fig. 10.9h and i).

10.7 Foraging around obstacles

Finding 17. *When plasmodium of* P. polycephalum *scouts a space with pseudopodia, it abandons the pseudopodia which encounter illuminated do-*

148   *Physarum machines*

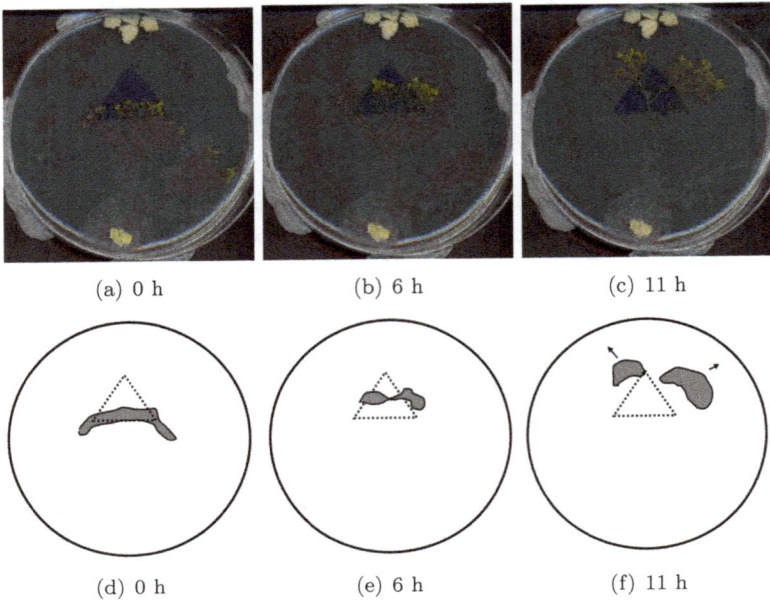

(a) 0 h   (b) 6 h   (c) 11 h

(d) 0 h   (e) 6 h   (f) 11 h

Fig. 10.8   Splitting plasmodium wave with illuminated triangle: (a)–(c) scanned images
of experimental Petri dish, (d)–(f) schemes of the propagating plasmodium.

*mains and sprouts new pseudopodia in the less illuminated areas and to-
wards the sources of nutrients.*

Nakagaki et al. [Nakagaki et al. (2007)] already demonstrated that the
plasmodium optimizes its protoplasmic network in the presence of light,
namely, the plasmodium reduces the number and size of protoplasmic tubes
exposed to light. We do not focus on optimization and do not provide statistical evidence but rather a few morphological examples of how a growing
protoplasmic network is shaped by illuminated domains.

The most typical example is shown in Fig. 10.10. Initially, an oat flake
colonized by the plasmodium is placed in the central part of the Petri dish
(Fig. 10.10a and c). Three intact oat flakes are positioned near the Petri
dish's wall so that they form vertices of a triangle centered at the initial
position of the plasmodium. The plasmodium sprouts several pseudopodia heading towards the oat flakes. Most of the pseudopodia encounter
illuminated areas and cease propagation. The pseudopodium propagating
east-south-east avoids illuminated areas and eventually reaches the north-
eastern oat flake (Fig. 10.10b and d).

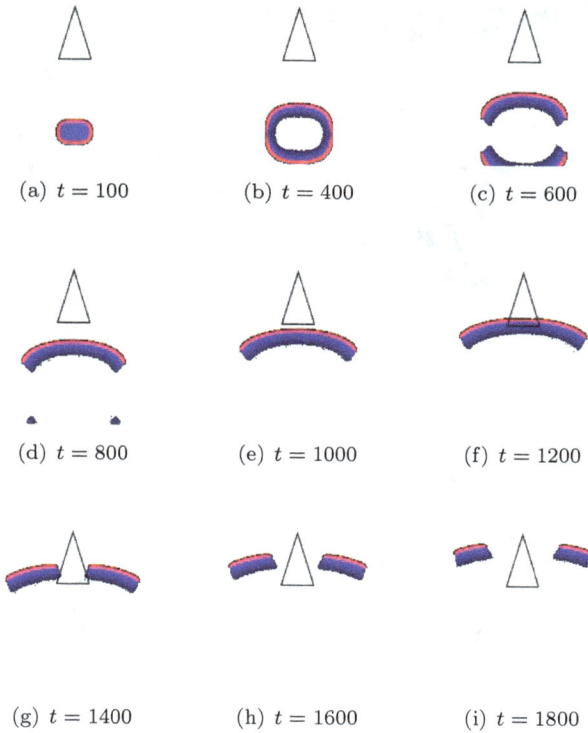

Fig. 10.9 Simulation of plasmodium wave split by illuminated triangle. Red and blue components of each pixel's color are defined as follows. If $u > 0.1$, then red value $255 \cdot u$, if $v > 0.1$, then blue value $600 \cdot v$, otherwise the background white color is used. Time shows steps of numerical integration.

There is evidence that the plasmodium may 'evaluate' the size of the light obstacles by sprouting pseudopodia in several directions, integrating information about the size of the obstacles (possibly via protoplasm oscillation frequencies) and then making a decision about where to sprout further pseudopodia. An example of such decision making is provided in Fig. 10.11. In the experiment reported, the plasmodium was placed in the center of a Petri dish, and additional oat flakes were placed in north and south poles of the dish. The plasmodium sprouts several pseudopodia to explore the space. Some of the pseudopodia encounter illuminated

(a) $t = 0$ h   (b) $t = 9$ h

(c) $t = 0$ h   (d) $t = 9$ h

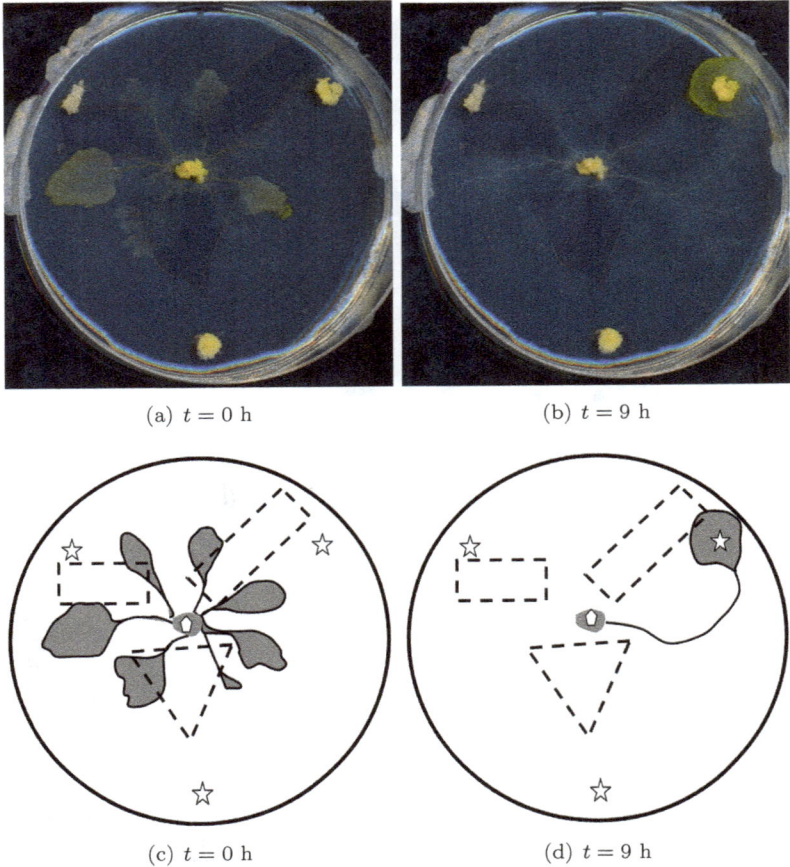

Fig. 10.10   Plasmodium scouting in arena with illumination obstacles: (a) and (b) experimental, (c) and (d) schemes of propagation. Oat flakes are shown by stars.

areas. The illuminated area south of the plasmodium is larger than the illuminated area north of the plasmodium. Therefore, the plasmodium does not sprout any more pseudopodia southward but produces a protoplasmic branch growing north. This pseudopodium reaches the south source of nutrients (Fig. 10.11).

Figure 10.12 shows a history of plasmodium probing a space by sprouting pseudopodia. Places where exploration is abandoned are marked by solid black disks in Fig. 10.12c. Scanning the scheme (Fig. 10.12c) from the south to the north, we see that the plasmodium probes the illuminated tri-

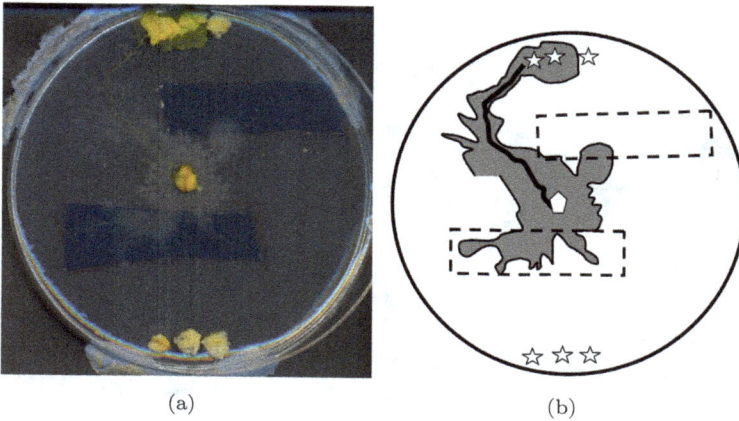

(a)   (b)

Fig. 10.11   Plasmodium scouting in arena with illumination obstacles: (a) snapshot of experiments, (b) scheme of propagation. Oat flakes are shown by stars.

angle and walls of the Petri dish (first two disks), then the wall and a space between the illuminated rectangle and triangle (further two disks). Final probing activity (three northern black disks in Fig. 10.12c) is associated with the plasmodium's close approach to the source of chemo-attractants.

Such a mechanism of scouting allows plasmodium to pre-shape its foraging network. When eventually the plasmodium spans all sources of nutrients with its protoplasmic network, major tubes of the network become positioned in the non-illuminated areas (Fig. 10.13).

10.8   Routing signals in Physarum machine

In Chap. 8.4, we experimentally demonstrated how active zones can be manipulated by dynamical addition of nutrients. Programming with nutrients is not really efficient, because once the source of nutrients is placed in the computing space, it irreversibly changes the configuration of attracting fields. Light inputs allow for on-line reconfiguration of obstacles and thus provide more opportunities for embedding complex programs in Physarum machines.

In the present chapter we provided experimental implementation of four operations with active zones (Fig. 10.14):

- ERASE,
- RIGHT,

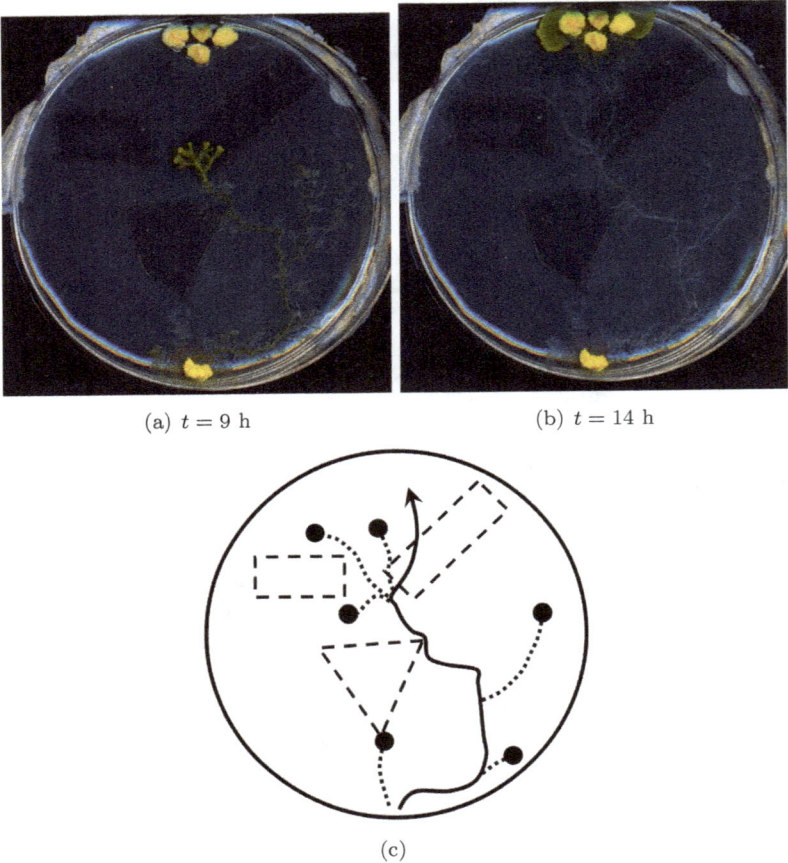

(a) $t = 9$ h   (b) $t = 14$ h

(c)

Fig. 10.12   Plasmodium scouting in arena with illumination obstacles: (a) and (b) snap-shots of experiments, (c) scheme of propagation.

- LEFT,
- MULTIPLY.

Operation ERASE(A) removes, or permanently deactivates, active zone A (Fig. 10.14a). The operation is implemented by placing a large enough — so that there is no chance of the plasmodium wave escaping from the shaded area – domain of illumination in front of the traveling zone A. The plasmodium wave A then gradually disappears. The operation is exempli-fied in Fig. 10.10, where a few active zones are canceled by rectangles of illumination.

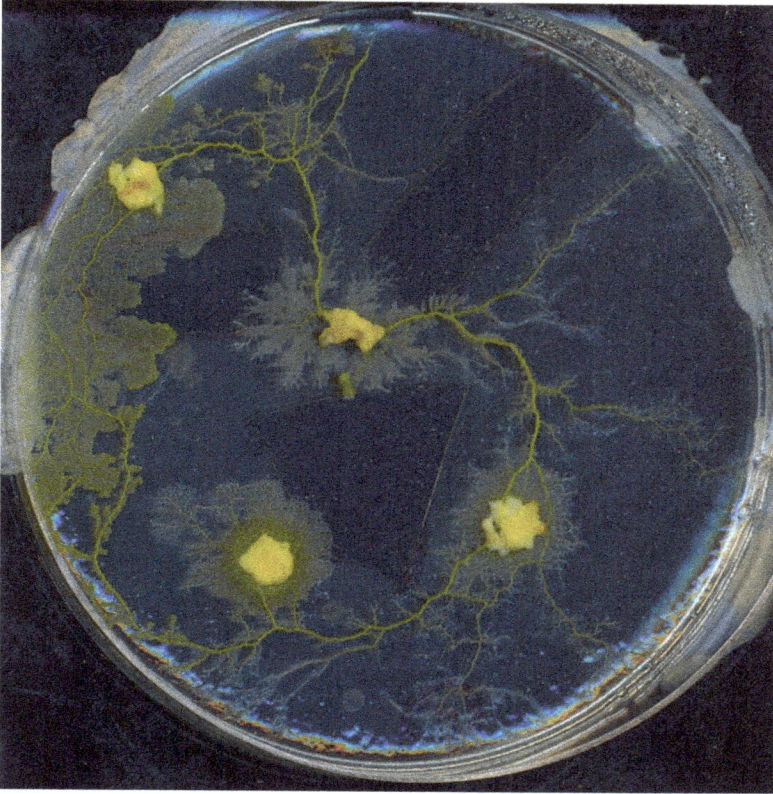

Fig. 10.13   Protoplasmic network shaped by light obstacles.

Operation LEFT(A) rotates the velocity vector of zone A by the angle α, $-90 \leq \alpha < 0$; for RIGHT(A), $0 < \alpha \leq 90$ (Fig. 10.14b and c). The exact value of the rotation angle α depends on several factors, including size of illumination domain, size of traveling zone A, humidity, overall illumination and fitness of the plasmodium. Based on a few successful experiments on rotation of plasmodium waves, we could say that $\alpha = 45° \pm 15°$.

Operation MULTIPLY(A) = $\{A_L, A_R\}$ (Fig. 10.14b and c) splits a traveling plasmodium wave A into zones A_L and A_R deviated slightly west and east compared to the original direction of A's travel.

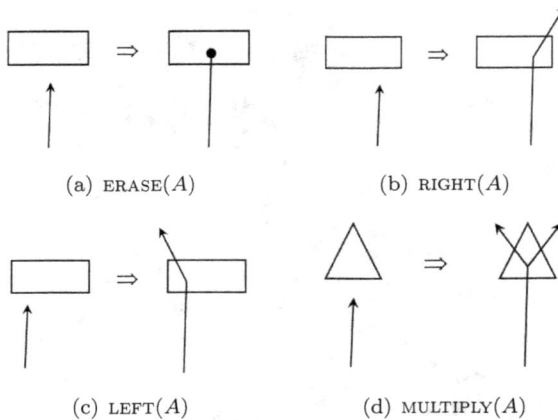

(a) ERASE(A)   (b) RIGHT(A)

(c) LEFT(A)   (d) MULTIPLY(A)

Fig. 10.14 Scheme of routing operations in Physarum machine. Directions of active zone movement are shown by arrows. Illuminated domains are rectangular and triangular shapes.

10.9 Disobedience

Classical experiments on Physarum photo-avoidance [Nakagaki et al (1999); Nakagaki et al. (2007)] use very strong, we could say eye-blinding, halogen lamps to negatively stimulate plasmodium. Any creature would try to escape from such strong illumination. In contrast, we use a rather 'gentle' dosage of light to fine tune plasmodium's behavior. The light intensity is low enough for plasmodium to attempt to ignore it.

The configuration of abandoned protoplasmic tubes in Fig. 10.15 tells a story of plasmodium 'stubbornness'. An oat flake colonized by plasmodium is inoculated in the south pole of the Petri dish. A few virgin oat flakes are placed in the north pole of the dish. They act as a source of chemo-attractants. There are two illuminated rectangles in the experimental arena. At first, the plasmodium tries to ignore the light and travels directly north by crossing the first illuminated rectangle. It abandons its route when stuck in the middle of the second illuminated rectangle (solid black disk in (Fig. 10.15b). The plasmodium 'recalculates' the route by crossing the first illuminated rectangle perpendicularly to the longest side and by avoiding the second illuminated rectangle (Fig. 10.15b).

In the experiments (Figs. 10.16 and 10.17), plasmodium crossed through the first illuminated obstacle, changed its trajectory and avoided further illuminated obstacles (Figs. 10.16b and Fig. 10.17c). In the experiment

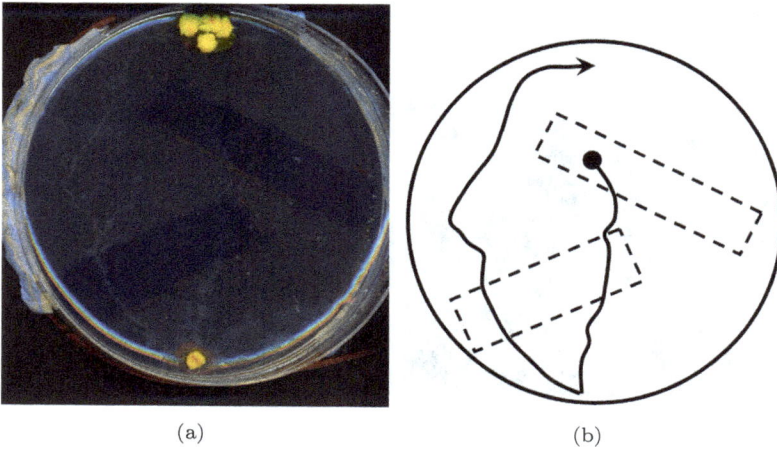

(a) (b)

Fig. 10.15   Scouting arena with illuminated obstacles: (a) scanned image of experimental Petri dish, (b) scheme of propagating plasmodium.

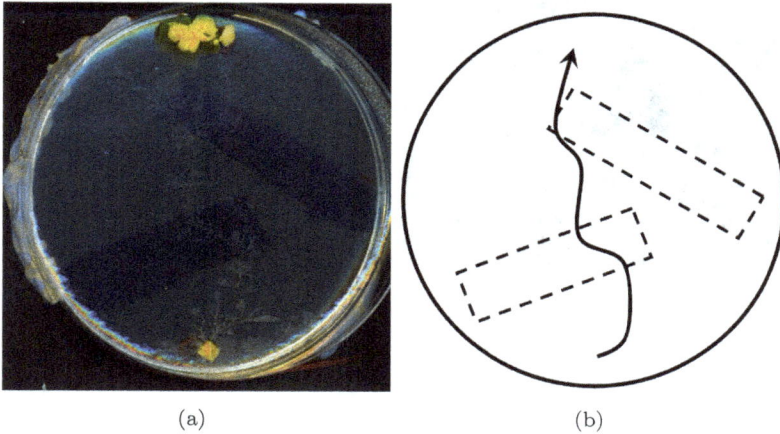

(a) (b)

Fig. 10.16   Scouting arena with illuminated obstacles: (a) scanned image of experimental Petri dish, (b) scheme of propagating plasmodium.

shown in Fig. 10.18, plasmodium probed (black disk in Fig. 10.18b) and avoided the illuminated triangle but crossed the illuminated rectangle (possibly because the plasmodium was very close to oat flakes).

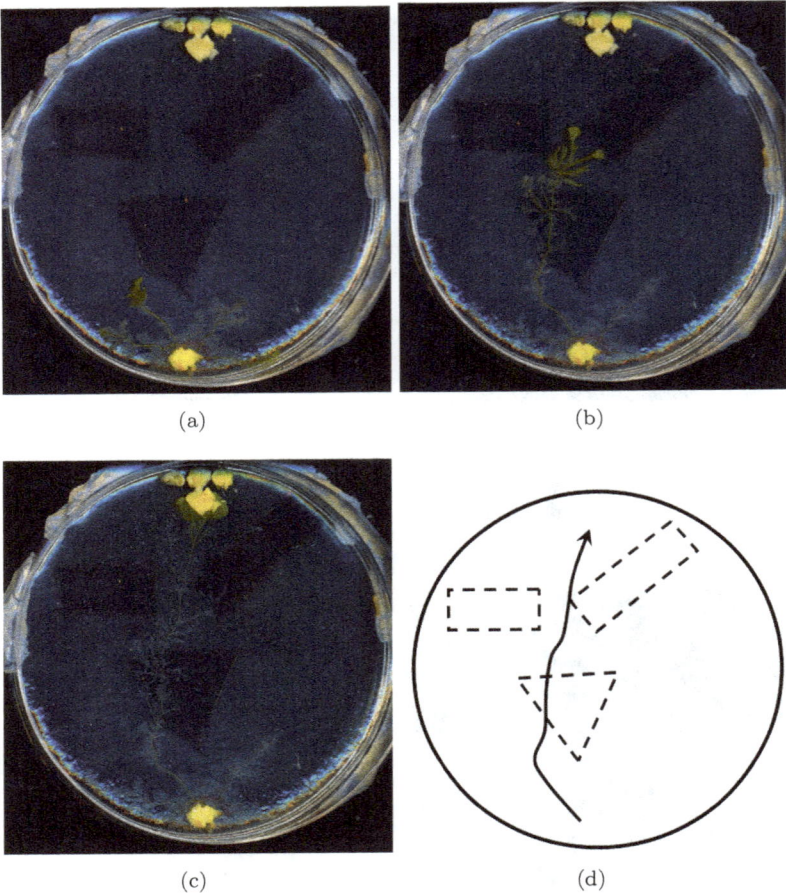

Fig. 10.17 Scouting arena with illuminated obstacles: (a)–(c) scanned images of experimental Petri dish, (d) scheme of propagating plasmodium.

10.10 Summary

In laboratory experiments with plasmodium of *P. polycephalum* and in computer simulations we discovered that propagating plasmodium interacts with illuminated domains similarly to traveling wave fragments in excitable media. We demonstrated that plasmodium waves can be diverted, annihilated and split by properly arranged shapes of illumination. The light-induced diversion of the plasmodium waves can be used as operations

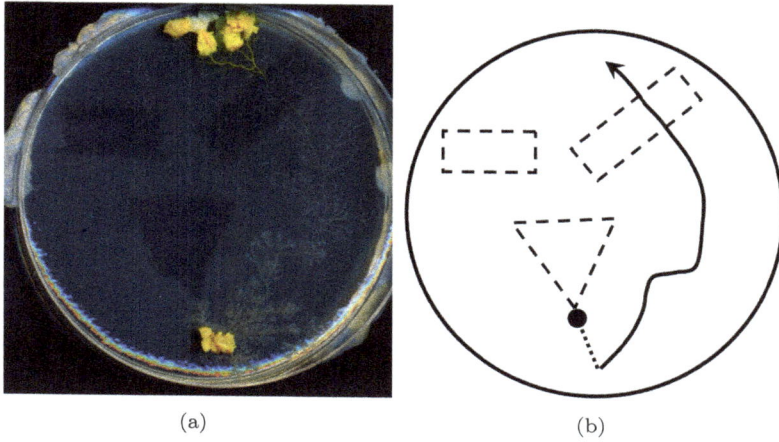

Fig. 10.18 Scouting arena with illuminated obstacles: (a) scanned image of experimental Petri dish, (b) scheme of propagating plasmodium.

on signals, or even as routing of mobile processes, in plasmodium-based implementations of general-purpose storage machines [Adamatzky (2007)].

Usually, a quite high intensity of illumination is used in studies of *P. polycephalum* response [Block and Wohlfarth-Bottermann (1981); Nakagaki et al. (2007)]. Thus, in [Block and Wohlfarth-Bottermann (1981)], the reported threshold intensity for blue light is about 1500 Lux. In our experiments the plasmodium was controlled by much less illuminated shapes, with a maximum intensity of illumination of 50–70 Lux. This is why we did not observe strong photo-avoidance reactions but rather gentle and often subtle changes in *P. polycephalum* behavior. The moderate intensity of illumination used in our experiments shows that we can achieve purposeful behavior of the plasmodium with light stimulation without causing any adverse reactions.

Geometrically shaped domains of illumination provided sharp boundaries between illuminated and non-illuminated areas. They did not form smooth gradients of illumination. Also, the size of the illuminated domains was the same order as a 'wavelength' of the propagating plasmodium waves. These guaranteed that the illuminated shapes acted rather as reflectors, or mirrors, than as sources of repelling fields. We envisage that the findings presented in the chapter can be used in experimental implementation of collision-based computing schemes [Adamatzky (2003)], where plasmodium waves represent quanta of information and illumination domains are used to route momentary wires, trajectories of the information quanta.

Chapter 11

Routing Physarum with repellents

This chapter is about controlling active zones of a Physarum machine. When a Kolmogorov–Uspensky machine [Kolmogorov (1953); Uspensky (1992)] is implemented in *P. polycephalum*, active zones are represented by localized propagating parts of plasmodium: pseudopodia, clusters of pseudopodia or localized wave fragments (Fig. 11.1).

To program the Physarum machine implementation of storage modification machines, we must physically control propagation of the active zones. We can route active zones with attracting or repelling fields, or a combination of them, similarly to methods of potential fields studied theo-

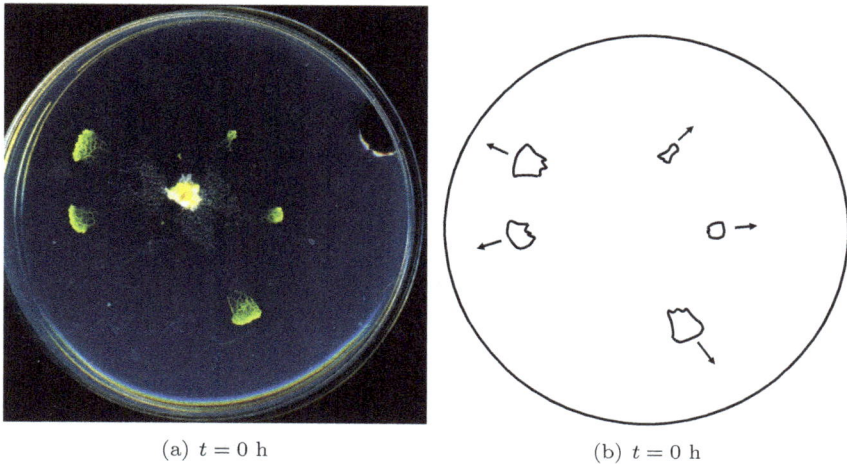

(a) $t = 0$ h   (b) $t = 0$ h

Fig. 11.1   Example of five active zones of Physarum machine traveling from the center of Petri dish: (a) experimental image of the dish with plasmodium, (b) scheme of the propagation.

retically and applied in robotics [Tarassenkp and Blake (1991)], artificial
intelligence [Hagelback and Johansson (2009)] and communication network
packet routing [Cheng et al. (1991)].

Routing and reconfiguration of plasmodium's active zones, or signals,
with attracting fields (sources of nutrients) is discussed in Chap. 8.4. In
laboratory experiments, the following operations are executed using attracting fields: merging of two active zones, multiplication of an active
zone, translation of an active zone from one data site to another and directing an active zone along a specified vector. In Chap. 9.4, we showed
that plasmodium's active zones can be, in principle, routed with domains of
localized illumination: the operations of deflection, splitting/multiplication
and erasing of active zones are achievable. However, light-based control
excels mainly in shaping of protoplasmic networks, not precise directing of
active zone propagation. Experimental evidence, see Chap. 9.4, shows that
active zones can respond to domains of illumination with long delays or —
depending on the plasmodium's physiological state — not respond at all.
We decide therefore to test another way of controlling active zones — with
chemo-repellents applied directly to the substrate.

In this chapter, we show how to control plasmodium with salt. We make
experiments within glass Petri dishes, diameter 9 cm. We use 2% agar
gel as a non-nutrient substrate and 2% corn meal agar gel as a nutrient-
rich substrate. To control plasmodium's propagation by repelling fields,
we use grains of coarse sea salt[1]. The Petri dishes are kept in the dark,
at temperatures of 22–25°C, except for observation and image recording.
Periodically, the dishes are scanned using an Epson Perfection 4490 scanner.

In Chap. 4.3, we have shown how to simulate space–time dynamics
of plasmodium using a two-variable Oregonator model. The simulation
is based on similarity between propagating pseudopodia and Belousov–
Zhabotinsky excitation waves. We use the model again in the present
chapter. In computational experiments we perturb the medium with an
initial excitation, where 30–50 sites in a chain are assigned value $u = 1.0$
each. The perturbation generates an elliptic excitation wave, which later
splits into two wave fragments. We are controlling one of the fragments.
We use grids of 100×100 and 150×150 sites. Obstacles induced by sodium
chloride crystals are simulated as fixed-radius disks of grid sites with values
$u = 0$ and $v = 1$.

---

[1]Saxa coarse grain sea salt, RHM Foods, Middlewich, CW10 0HD, UK.

(a) $t = 0$ h

(b) $t = 4$ h

(c) $t = 8$ h

(d) $t = 24$ h

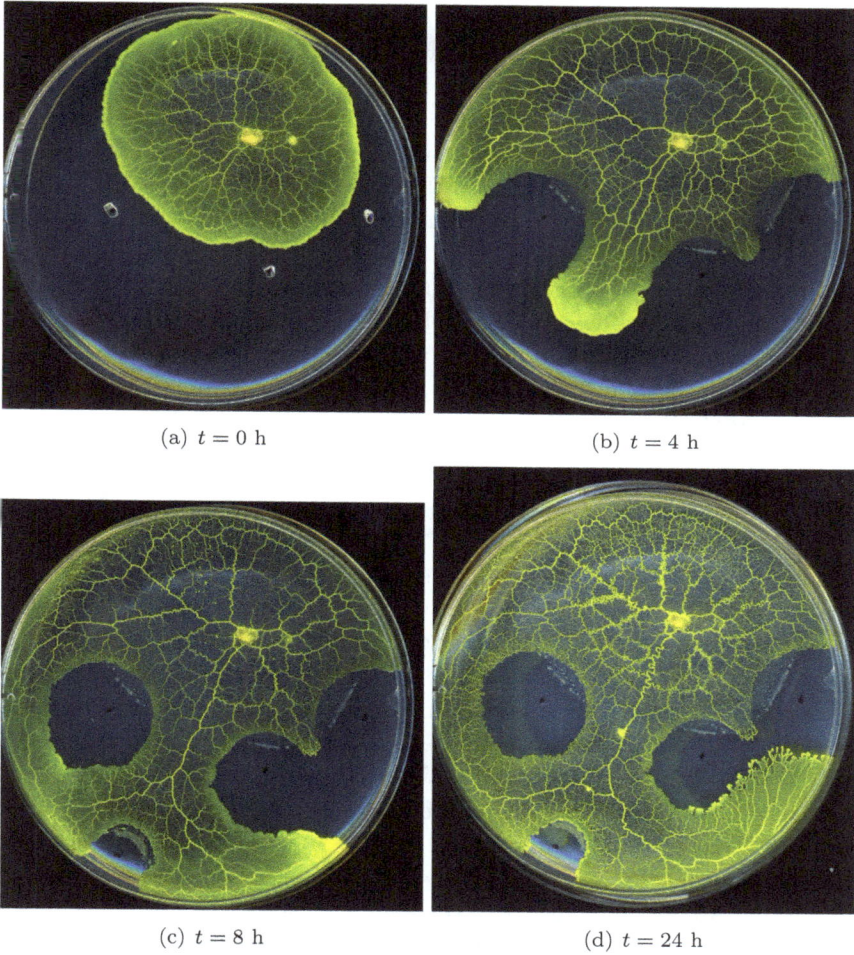

Fig. 11.2 Plasmodium avoids salted domains while propagating on a nutrient-rich substrate. These are scanned images of experimental arena with positions of salt grains marked by black dots on the bottom of Petri dish. South-western crystal was placed on gel at step (b).

11.1 Avoiding repellents on nutrient-rich substrate

On a nutrient-rich substrate a plasmodium wave propagates around domains with a high concentration of sodium chloride as if they were impenetrable obstacles (Fig. 11.2). Initially, the plasmodium propagates in a circular shape (Fig. 11.2a). On encountering an increased concentration of salt, the plasmodium forms two growth regions (Fig. 11.2b). The growth regions, which exhibit a higher rate of metabolism, are highlighted in Fig. 11.3a as domains where red and green values of color differ by at most 30 units. The active zones 'squeeze' between obstacles and expand after leaving the gaps (Fig. 11.2c). Eventually, all the substrate is colonized by plasmodium apart from domains of high concentration of sodium chloride (Fig. 11.2d).

Propagation of growing plasmodium fronts is satisfactorily reproduced in the Oregonator model; two examples of excitation wave fronts overlaid with the original plasmodium image are shown in Fig. 11.3b and c. The principal protoplasmic tubes coincide with sites of maximum curvature of the wave fronts (Fig. 11.3d).

11.2   Operating on non-nutrient substrate

Cultivating plasmodium on a nutrient-rich substrate does not bring substantial benefits from the unconventional computing point of view. On the nutrient-rich substrate, the Physarum machine can only send one-to-all signals and would require geometrical constraining to implement a logical circuit. On a non-nutrient substrate, the plasmodium propagates as a traveling localization.. The active zone of the plasmodium keeps its shape and size for a substantial period of time. The plasmodium's traveling localization is analogous to propagating wave fronts in a sub-excitable Belousov–Zhabotinsky medium [Adamatzky et al. (2008)]. It can be used in collision-based computing circuits [Adamatzky (2003)]. In terms of collision-based computing [Adamatzky (2003)], a traveling plasmodium localization, or an active zone, is interpreted as a variable, or a signal, whose value is coded in the propagation vector.

Using repelling fields generated by crystals of sodium chloride, one can implement the following operations over plasmodium's active zones A and B:

- DEFLECT(A): change direction of active zone A's movement,
- MULTIPLY(A): split active zone A into two independent active zones A_1 and A_2 of like size and shape,
- MERGE(A, B): merge two active zones A and B into one active zone.

Schemes of the operations with active zones, or propagating signals, are

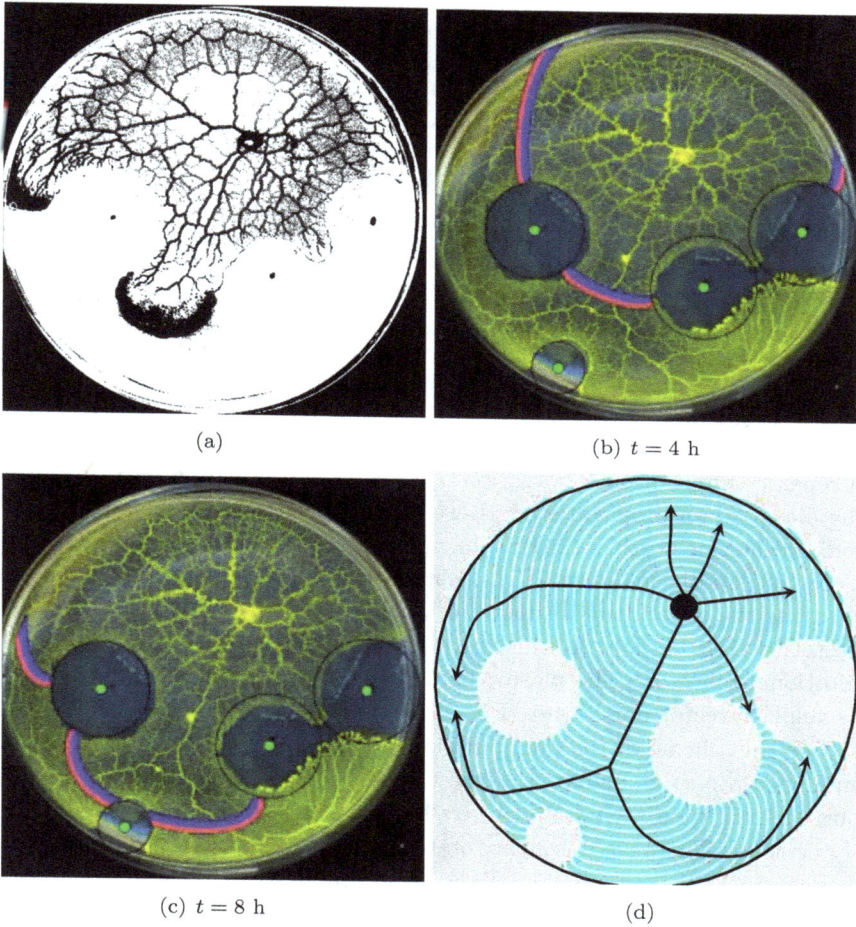

(a)

(b) $t = 4$ h

(c) $t = 8$ h

(d)

Fig. 11.3   Analyzing data from Fig. 11.2: (a) regions of active growth are highlighted, (b) and (c) snapshots of simulated propagation of wave front in Oregonator model overlaid with image (Fig. 11.2d) of plasmodium, (d) time-lapse contour of wave fronts in Oregonator model overlaid with sketch of main protoplasmic tubes from Fig. 11.2d.

shown in Fig. 11.4. We now provide experimental evidence of the above claims.

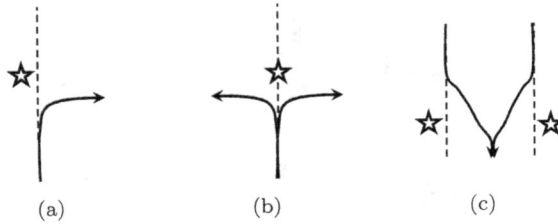

Fig. 11.4 Basic routing operations over plasmodium's active zones A and B: (a) DEFLECT(A), (b) MULTIPLY(A), (c) MERGE(A, B). Trajectories of traveling active zones, or signals, are shown by solid arrows, positions of repellent sources by stars. Trajectories of signals in absence of repellents are shown by dotted lines.

11.3 Operation DEFLECT

A repeated application of operation DEFLECT is shown in Fig. 11.5. Initially, the plasmodium propagates south-west (Fig. 11.5a). We place a crystal of sodium chloride north of the traveling plasmodium. The plasmodium makes a 45° turn and heads south (Fig. 11.5b). By adding two additional crystals of salt to the experimental substrate, we cause the plasmodium to turn 90° westward (Fig. 11.5c). At this stage the wall of the Petri dish acts as an additional reflector. The plasmodium is deflected by the wall and orients its velocity vector north-west (Fig. 11.5d).

See the scheme of the manipulation in Fig. 11.6a. These experimental findings are partly simulated in the Oregonator model (Fig. 11.6b–e). To turn an excitation wave fragment by 180°, we arrange a semicircular chain of obstacles. The wave fragment rotates by approx. 45° after collision with each obstacle (Fig. 11.6b–e). Time-lapse images of wave fronts and the trajectory of the wave fragment motion are shown in Fig. 11.6f.

11.4 Operation MULTIPLY

Experimental and simulated implementations of operation MULTIPLY are illustrated in Fig. 11.7. There are two oat flakes in the snapshots (Fig. 11.7a and b) of the experimental system. Initially, the plasmodium is inoculated in the eastern flake, shown by the pentagon in Fig. 11.7c and d. The western oat flake (rectangle in Fig. 11.7c and d) is used as an auxiliary attracting field to determine the vector of the initial propagation of the plasmodium. The plasmodium propagates westward. When the plasmodium reaches the attracting oat flake (Fig. 11.7a), we place two crystals of sodium chloride,

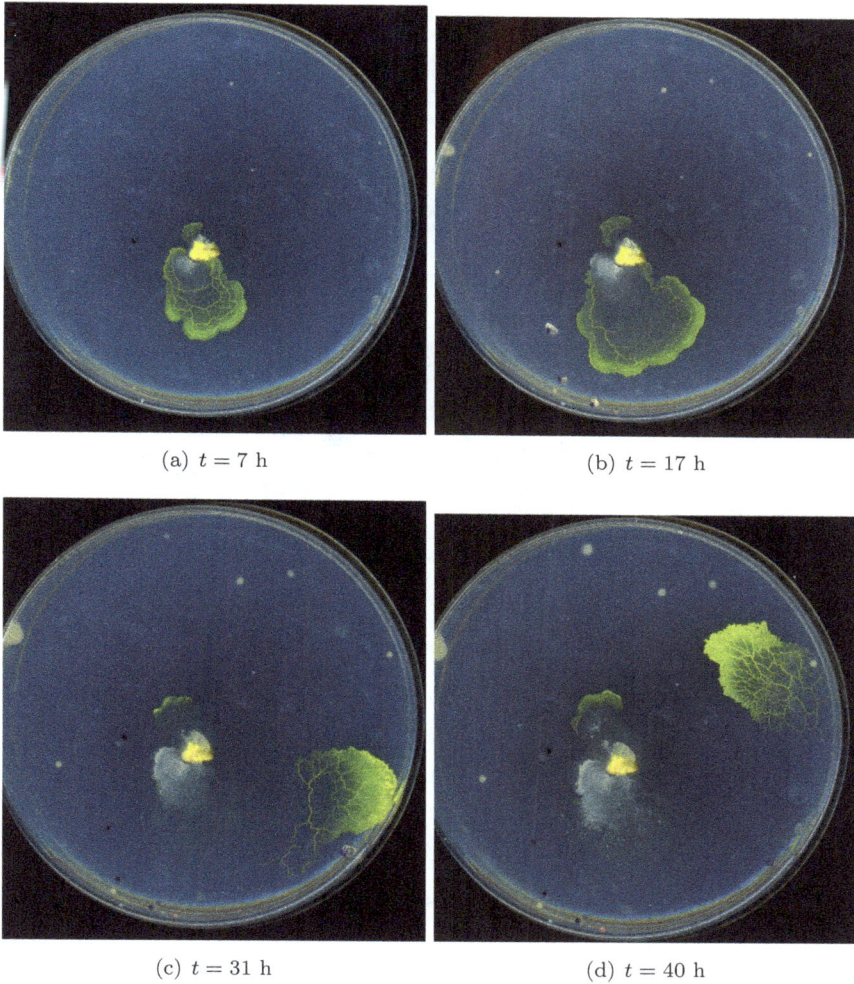

(a) $t = 7$ h

(b) $t = 17$ h

(c) $t = 31$ h

(d) $t = 40$ h

Fig. 11.5 Implementation of operation DEFLECT: turning plasmodium wave anti-clockwise. Experimental images of the propagating plasmodium.

shown by shaded stars in Fig. 11.7c. Due to repelling fields formed by the crystals, the plasmodium does not propagate forwards. Instead, the plasmodium sub-divides itself into two propagating localizations. One localization travels north-west and the other south-east (Fig. 11.7b and d).

The operation MULTIPLY is satisfactorily simulated in the Oregonator model; see Fig. 11.8a and d. We generate a wave fragment which travels

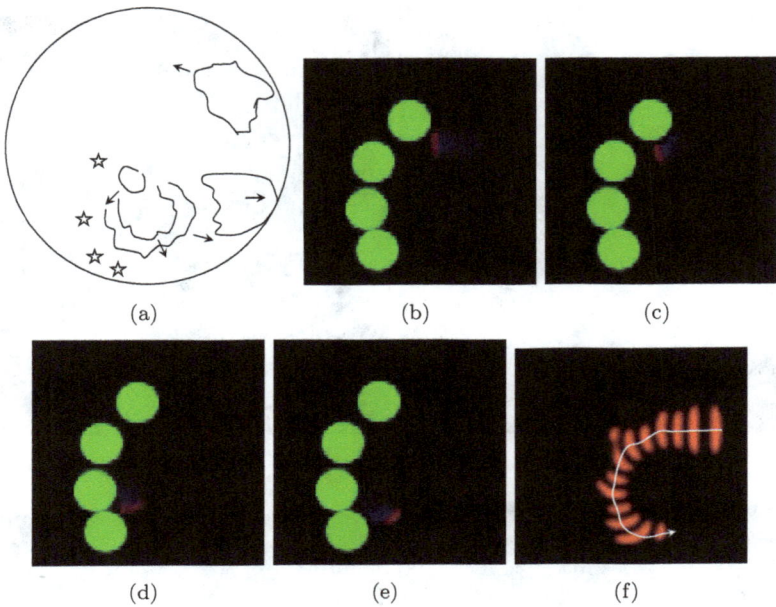

Fig. 11.6 Implementation of operation DEFLECT: turning plasmodium wave anti-clockwise: (a) scheme of the active zones controlled by repelling fields, as shown in Fig. 11.5: plasmodium is pictured as closed shapes with velocity vectors, salt crystals are shown by stars, (b)–(e) snapshots of wave propagating in Oregonator model, repelling obstacles are colored in green, (f) overlaid images of the propagating and rotating wave front in Oregonator model.

west (Fig. 11.8a) and arrange a configuration of repelling fields ahead of it. The wave fragment collides with the repelling fields (Fig. 11.8b) and splits into two wave fragments: one fragment travels north and the other south (Fig. 11.8c and d). Time-lapse contours of the wave fragments are shown in Fig. 11.8e.

Finding 18. *The operations* DEFLECT *and* MULTIPLY, *over the plasmodium's active zone(s), can be implemented with one source of repelling fields.*

Finding 19. *The operation* MERGE *can be executed only with at least two sources of repelling fields.*

(a) $t = 0$ h

(b) $t = 4$ h

(c) $t = 0$ h

(d) $t = 4$ h

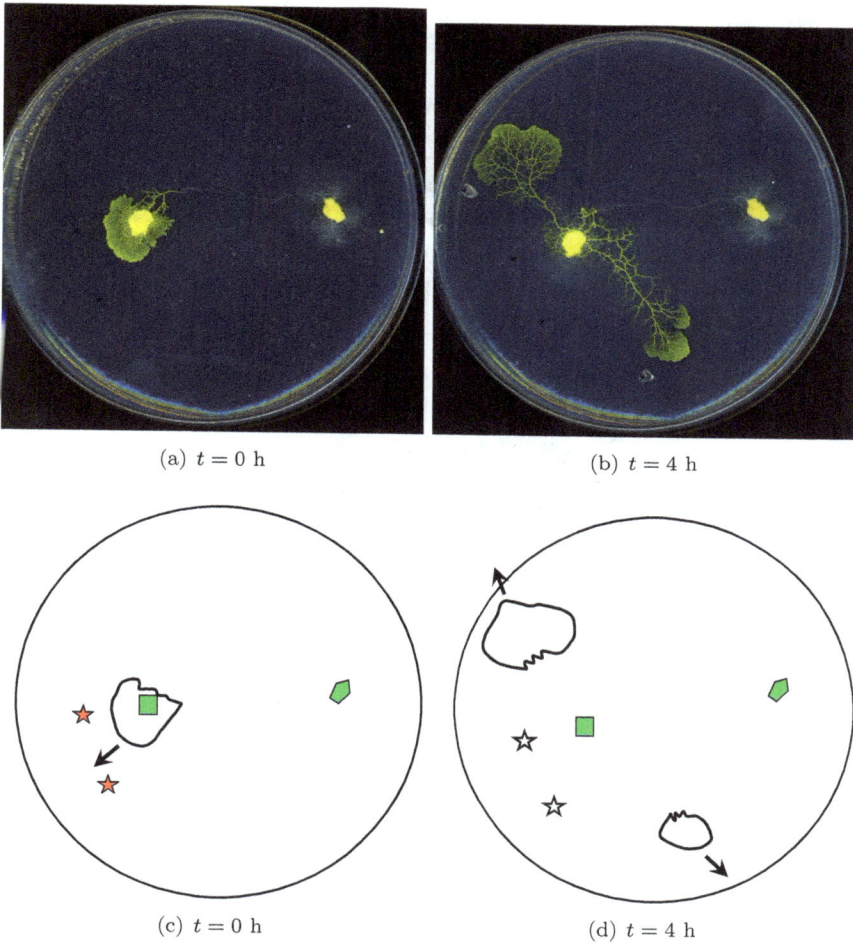

Fig. 11.7   Implementation of operation MULTIPLY: (a) and (b) scanned images of propagating plasmodium, (c) and (d) scheme of the manipulation: original position of plasmodium is shown by pentagon, attracting oat flake by square, locations of salt crystals by stars, shaded stars indicate salt crystals placed at the moment of image scanning.

11.5   Operation MERGE

We can assume that two plasmodium localizations A and B propagate with the same speed, far from walls of the experimental container, and their initial velocity vectors are parallel or oriented outwards from each other. By having just one source of repellent, we can deflect either one

(a)   (b)   (c)

(d)   (e)

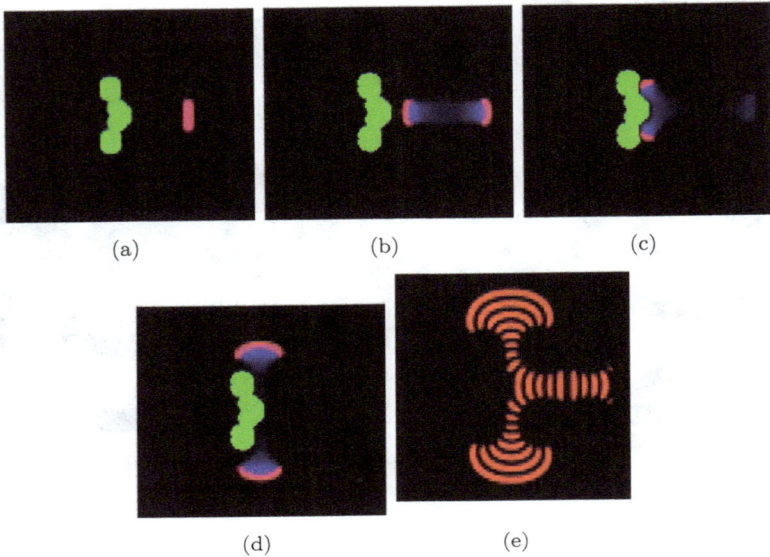

Fig. 11.8 Implementation of operation MULTIPLY: (a)–(d) snapshots of Oregonator model, showing initial disposition of salt crystals and plasmodium (a), propagation of plasmodium active zone towards the crystals (b), and splitting of the active zone into two active zones (c) and (d); (e) time-lapse images of propagating wave front. Repellent domains, generated by salt crystals, are shown by solid green disks.

of the localizations or both localizations at the same time. To deflect the localizations and orient their velocity vectors towards each other, one must place at least two sources of repellents — one source for localization A, the other for B.

Examples of experimental and simulated merging are shown in Figs. 11.9 and 11.10; see also scheme of the plasmodium's behavior in Fig. 11.11. An active zone propagating south-east encounters a repelling field, generated by a crystal of sodium chloride (Figs. 11.9a, 11.10b and c and 11.11a). The active zone splits into two zones. Two newly formed localizations are deflected, by the same repelling field, south-west (Figs. 11.9b, 11.10c and d and 11.11b). The localizations travel along parallel trajectories and slightly expand (Figs. 11.9c, 11.10e and f and 11.11c). We add two more salt grains to generate repelling fields on both sides of the propagating localizations. Being 'compressed' by the repelling fields, the active zones reorient their velocity vectors towards each other, collide with each other and merge (Figs. 11.9d, 11.10f and g and11.11d). The newly formed local-

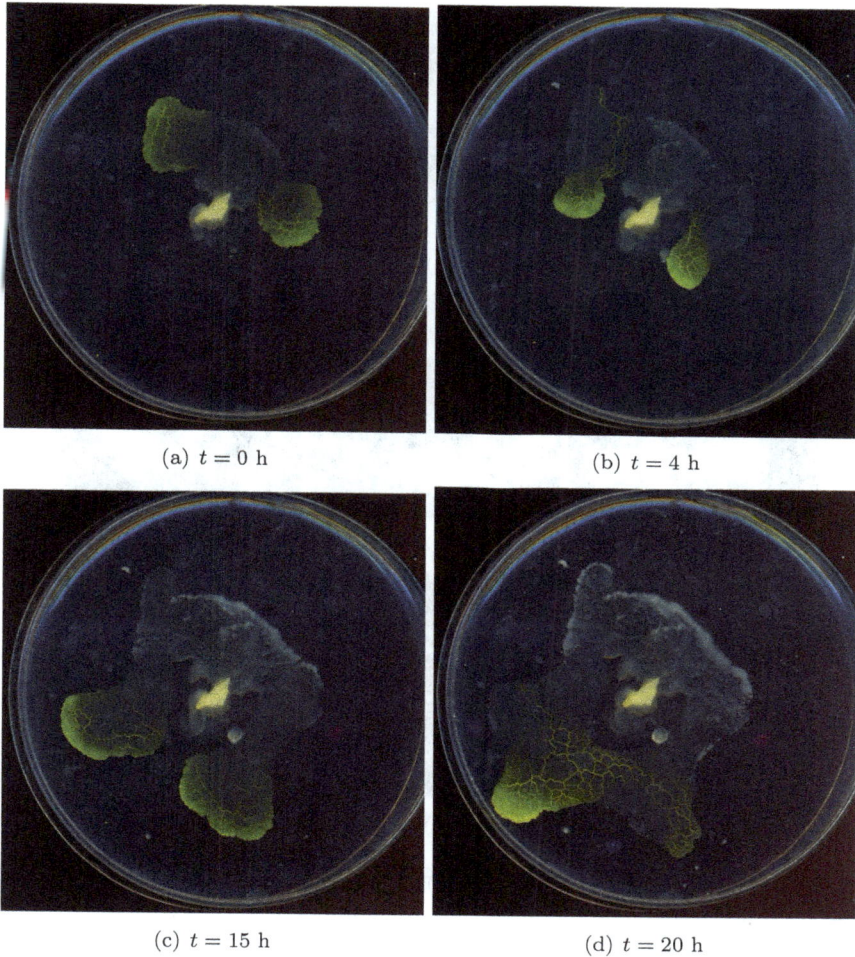

(a) $t = 0$ h

(b) $t = 4$ h

(c) $t = 15$ h

(d) $t = 20$ h

Fig. 11.9 Implementation of MULTIPLY (a), DEFLECT (b) and MERGE (c) and (d) operations. Snapshots of experimental container. Salt grains are marked by black disks on the bottom of Petri dish.

ization propagates south-west (Figs. 11.9d and 11.11).

Going back to the beginning of the chapter, have a look again at Fig. 11.1. We decided to merge two active zones propagating west-north-west (Fig. 11.1). We placed two crystals of sodium chloride, one north of the north-west-traveling localization, the other south of the west-traveling localization (Fig. 11.12a). The active zones sharply changed the directions

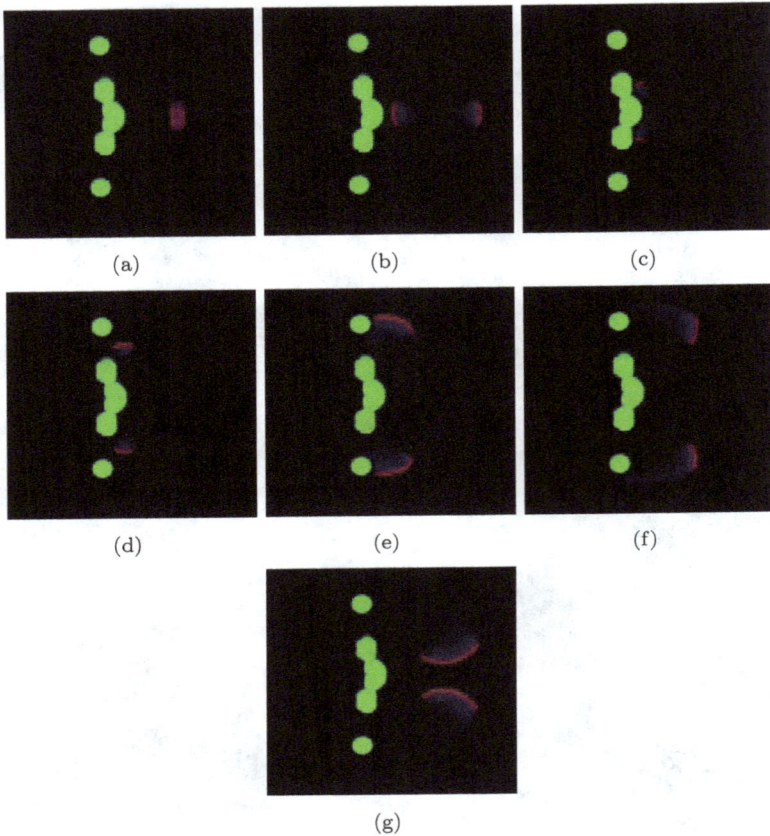

Fig. 11.10 Implementation of MULTIPLY (a)–(c), DEFLECT (d)–(f) and MERGE (g) operations. Snapshots of Oregonator model simulating the propagating plasmodium. Repellent domains generated by salt crystals are shown as solid green disks. Note that in this particular simulation setup we used top and bottom absorbing boundaries to aid rotating wave fragments towards each other.

of their motion (Fig. 11.12b), faced each other and merged (Fig. 11.12c).

11.6 Summary

We experimentally demonstrated how to control the localized propagation of plasmodium of *P. polycephalum* by generating repelling fields with sodium chloride. Experimental results are verified by numerical simulation

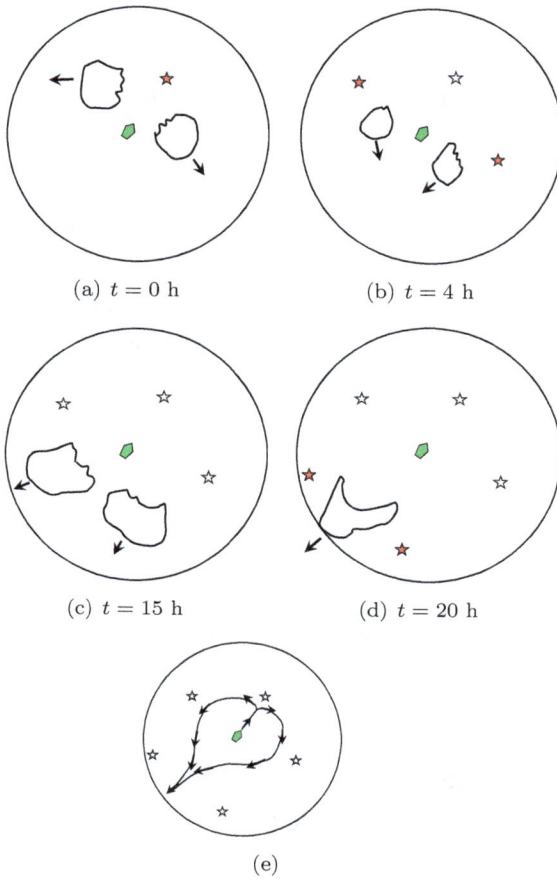

(a) $t = 0$ h

(b) $t = 4$ h

(c) $t = 15$ h

(d) $t = 20$ h

(e)

Fig. 11.11 Scheme of plasmodium's controlled behavior (a)–(d) experimentally illustrated in Fig. 11.9: positions of salt grains added at current time step are shown by filled stars, crystals added at previous steps are shown by non-filled stars. Initial position of plasmodium is shown by filled pentagon. Trajectories of propagating active zones are shown in (e).

of propagating localizations in the two-variable Oregonator model. Basic operations over propagating localizations, or active zones — DEFLECT, MULTIPLY and MERGE — are shown in Fig. 11.4. The results obtained fit well in the framework of controlling plasmodium with repelling (light), Chap. 9.4, and attracting (sources of nutrients), Chap. 8.4, fields.

All operations achievable by geometrically defined domains of illumination can also be implemented by an appropriate configuration of salt

(a) $t = 0$ h   (b) $t = 8$ h

(c) $t = 15$ h

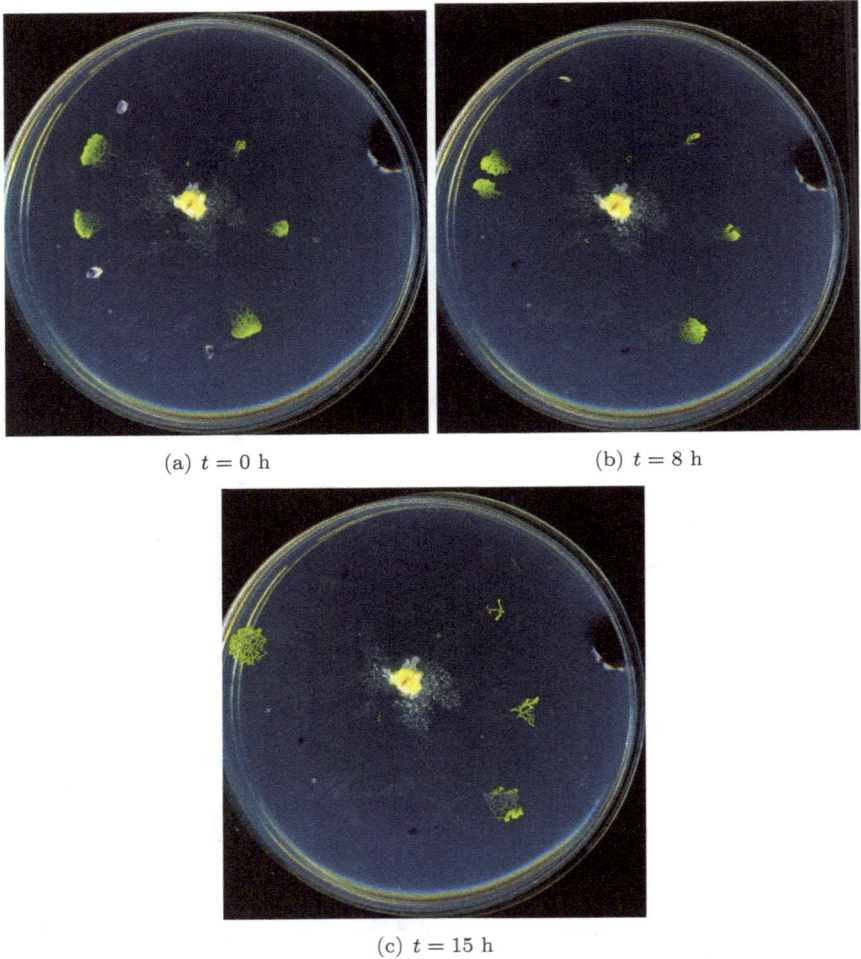

Fig. 11.12   Merging of active zones. Two grains of salt, clearly visible as white spots with black centers in the north-western sector of the Petri dish, are placed near active zones to be merged (a). The zones avoid a high concentration of salt, turn towards each other (b) and merge (c).

crystals. Outcomes of the control with salt crystals are more predictable than outcomes of control with gradients of illumination. However, we could not, in general, implement a dynamical programming of plasmodium be-havior with salt, because substrate domains with increased concentration of chemo-repellents could not be easily reused by plasmodium.

Comparing repellent- and attractant-based control techniques, we found that operations DEFLECT and MULTIPLY can also be executed using sources of nutrients; see Chap. 8.4. Operation MERGE is implementable only with repelling fields (light or salt).

In contrast to traditional methods of controlling active zones of *P. poly-cephalum*, which guide plasmodium at every stage of its propagation, our techniques allow active zones to be in 'free flight'. Most of the time, an active zone travels without any interaction with repelling fields. Only when we want to change the active zone's trajectory do we set up reflectors ahead of the propagating localization. Based on such an approach, we can adopt principles of collision-based computing [Adamatzky (2003)] and implement plasmodium-derived computing circuits. In such dynamical circuits, logical variables can be represented by propagating plasmodium localizations. Values of the variables are encoded in velocity vectors of the plasmodium's active zones. When an active zone encounters a repelling or attracting field, or another active zone, its velocity vector changes. The change in the velocity vector is reflected in the value update of the logical variable represented by the active zone.

Chapter 12

Physarum manipulators

Plasmodium is a reaction–diffusion, excitable, medium encapsulated in a membrane. Waves of excitation propagating inside the plasmodium are followed by waves of contraction. We know that chemical waves can transport lightweight objects. Chemical waves propagating in a BZ medium induce various types of convective flows in the liquid phase [Wu et al. (1995); Matthiessen and Muller (1996)]. Directions of the convective flows are determined by velocity vectors of propagating excitation wave fronts. The convective flows could, in principle, manipulate objects placed on a surface of the medium.

Experiments with a simulated massive array of closed-loop, i.e. sensing objects, actuators controlled by excitation dynamics in a BZ medium [Skachek et al. (2005); Adamatzky et al. (2005); Skachek et al. (2006)] showed that all types of traveling excitations, including target waves, spiral waves and wave fragments, can transport and manipulate small objects and planar shapes. An embedded reaction–diffusion controller for an open-loop, i.e. non-sensing manipulated objects, all-wet transporter is implemented by co-polymerizing ruthenium bipyridine with the volume-changing polymer isopropylacrylamide [Murase et al. (2009); Yoshida et al. (2009)]. The hybrid co-polymer is immersed in a catalyst-free BZ solution. The BZ reaction takes place in the gel. A propagating excitation wave front induces an associated mechanical wave of contraction due to the gel's response to charges of the ruthenium complex. Periodically excited waves, traveling along the gel, produce peristaltic motion, which transports lightweight objects placed on top of the gel sheet.

Can we employ the wave activity of plasmodium to transport lightweight objects on the plasmodium's surface similarly to transportation of objects by reaction–diffusion chemical waves? We tried to achieve such transportation in a few scoping experiments but failed. A lightweight object, e.g. a piece of foam, placed on a plasmodium membrane oscillates vertically but does not move laterally. However, we managed to achieve other kinds of manipulation by a plasmodium growing on a water surface.

To demonstrate that a biological substrate is suitable for robotic implementations, one must demonstrate that the substrate senses its environment, responds to external stimulus, solves complex computational tasks on spatially distributed datasets, moves itself and manipulates objects it is attached to. In [Adamatzky and Jones (2008)], we provided basic evidence that plasmodium of *P. polycephalum* can be a good biological prototype of future amorphous biological robots. We overview the results in the present chapter.

We used several experimental arenas to study behavior of the plasmodium and plasmodium-induced manipulation of floating objects. These are Petri dishes with base diameters 20 mm and 90 mm, and rectangular plastic containers of 200×150 mm^2. The dishes and containers were filled by 1/3 with distilled water. Data points, to be spanned by the plasmodium, were represented either by 5–10 mm sized pieces of plastic foam, which were either fixed to the bottom of the Petri dishes or left floating on the water surface, in the case of large containers. Oat flakes were placed on top of the foam pieces to program the plasmodium's behavior. Foam pieces, where plasmodium was initially placed, and the pieces with oat flakes were anchored to the bottom of the containers. Tiny foam pieces to be manipulated by plasmodium were left free floating.

12.1 Plasmodium on water surface

Plasmodium placed on a water surface propagates as on a non-nutrient substrate. It grows fine-branched trees of protoplasmic tubes, senses sources of nutrients and travels towards them. The plasmodium spans the nutrients with its protoplasmic network. The surface of the water is in tension. It physically supports propagating plasmodium if the plasmodium's contact weight to contact area ratio is small. When placed in an experimental container, the plasmodium forms pseudopodia in a search for sources of nutrients. In most experiments the 'growth part' of the pseudopodia has a tree-like structure for fine detection of chemo-gradients in the medium. Examples of tree-like propagating pseudopodia are shown in Fig. 12.1. The fine structure of the plasmodium tree also minimizes the weight to area ratio (Fig. 12.1a).

In many cases the plasmodium detects sources of nutrients and propagates towards them. In Fig. 12.1b, the plasmodium detected a group of oat flakes on the western floater and emitted a pseudopodium towards this source.

Pseudopodia do not always grow towards sources of nutrients. There is a pseudopodium growing south-west, where no sources of nutrients are located (Fig. 12.1b). This happens possibly because in large-sized containers the volume of air is too large to support a reliable and stationary gradient of chemo-attractants.

In Petri dishes the volume of air is relatively small and stationary. The plasmodium therefore easily locates sources of nutrients (Fig. 12.2). It builds spanning trees, where graph nodes, to be spanned, are represented by pieces of foam with oat flakes on top. Figure 12.2a shows that originally the plasmodium was positioned at the southern domain. In 12 h the plasmodium builds a link with the western domain, and then starts to propagate pseudopodia to the eastern domain (Fig. 12.2b).

When the plasmodium spans sources of nutrients, it produces many 'redundant' branches, i.e. additional links (Fig. 12.2). These branches of pseudopodia are necessary for space exploration but do not represent minimal edges connecting the nodes of the spanning tree. These 'redundant' branches are removed at later stages of the plasmodium spanning tree development. See a well-established spanning tree, a chain in this particular case, of data points in Fig. 12.3. Initially, the plasmodium was placed in the western domain. The plasmodium constructed the spanning tree in 15 h.

The plasmodium explores space and spans food sources with a protoplasmic tree, when placed initially on one of the floating objects. Would the plasmodium remain operational when placed just on the surface of the water without any solid surface support? Yes, Fig. 12.4 shows that the plasmodium does it perfectly. We placed a piece of plasmodium on a bare surface of water (Fig. 12.4a). In 3 h the plasmodium forms an almost circular front of propagating pseudopodia (Fig. 12.4b). In 8 h the front reaches two stationary domains with oat flakes (Fig. 12.4c and d).

Finding 20. *Plasmodium satisfactorily grows on a water surface, detects and colonizes floating sources of nutrients and develops protoplasmic networks.*

(a)

(b)

Fig. 12.1 The plasmodium explores experimental arena by propagating tree-like pseudopodia: (a) fine structure of the protoplasmic tree, (b) plasmodium senses a source of nutrients and propagates towards the source.

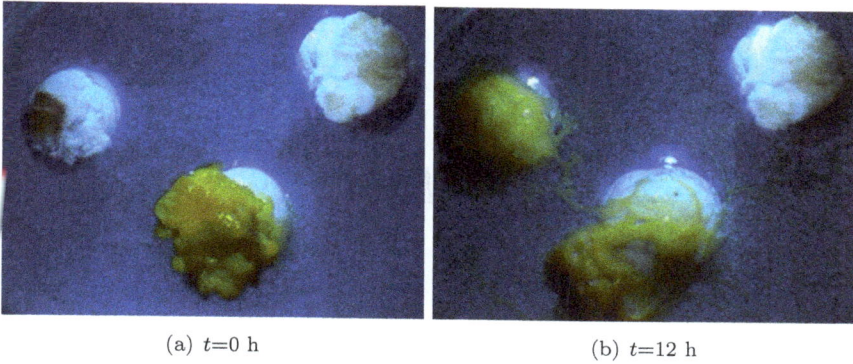

(a) $t=0$ h   (b) $t=12$ h

Fig. 12.2 Plasmodium builds links connecting its original domain with two new sites. ×10 zoom.

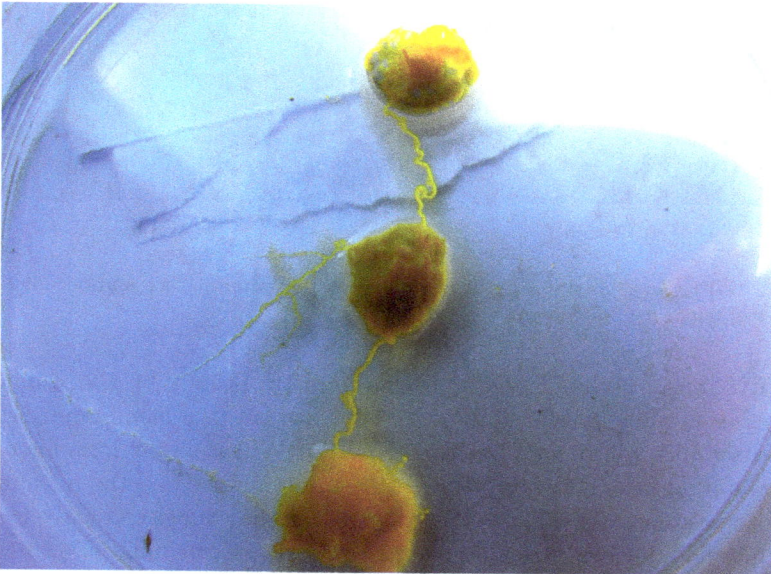

Fig. 12.3 Spanning tree of three points constructed by the plasmodium.

12.2 Manipulating floating objects

In usual conditions — on a solid or gel substrate — edges of spanning trees, represented by protoplasmic tubes, stick to the surface of the substrate. Therefore, the edges cannot move sideways. The plasmodium can

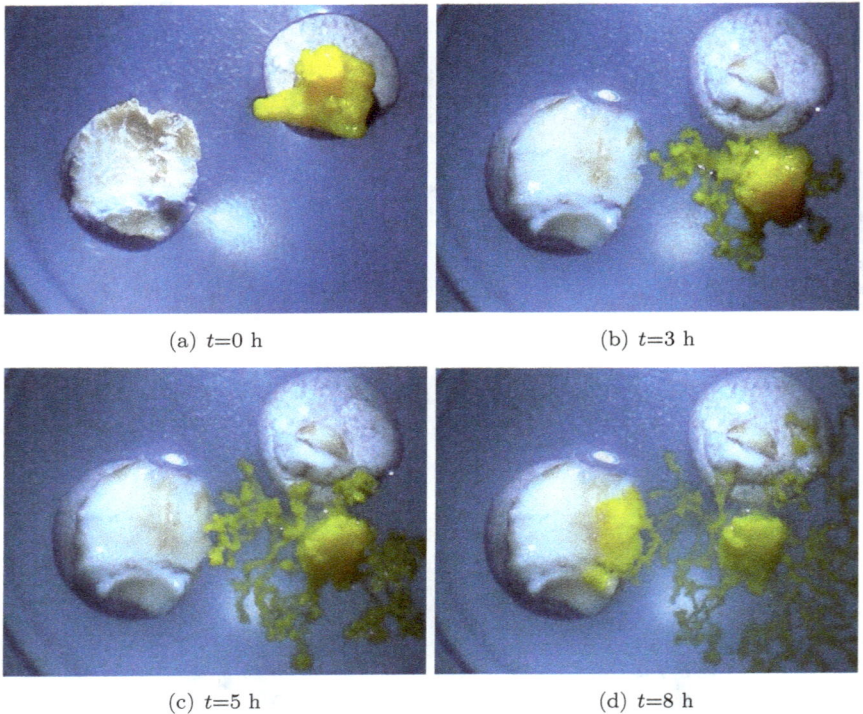

(a) $t=0$ h   (b) $t=3$ h

(c) $t=5$ h   (d) $t=8$ h

Fig. 12.4   Plasmodium starts its development on the water surface and occupies two sources of nutrients. ×10 zoom.

do a dynamical update of its protoplasmic network only by abandoning a protoplasmic tube and forming a new tube instead (the membrane shell of the ceased link will remain on the substrate). When plasmodium operates on a water surface, cohesion between the water surface and the membrane of protoplasmic tubes is small. So, the protoplasmic tubes can move freely. The plasmodium can straighten up its tubes and can minimize costs of transfer and communication between its distant parts.

Straightening up of the protoplasmic tubes is illustrated in Fig. 12.5: the tubes contract and thus become shorter.

Evidence of tube contraction leads us to the suggestion that if two floating objects, both with sources of nutrients, are connected by a protoplasmic tube, then the objects will be pulled together due to shortening of the protoplasmic tube. We did not manage to demonstrate the exact phenomenon of pulling two floating objects together; however, we found experimental

(a) t=0 h

(b) t=12 h

(c) t=0 h

(d) t=12 h

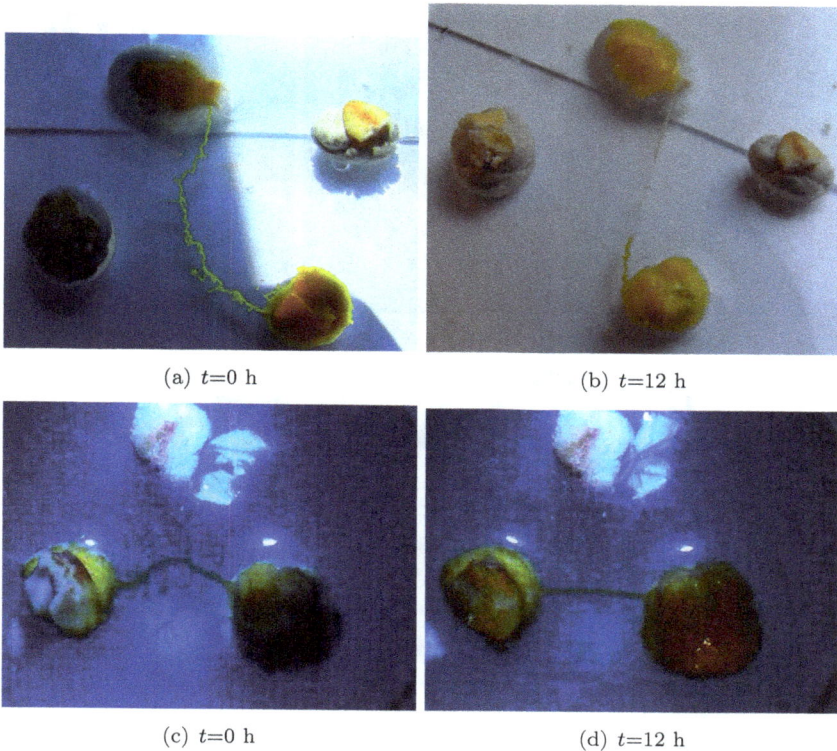

Fig. 12.5   Examples of straightening up of protoplasmic tubes: (a) and (c) plasmodium just formed a tube connecting two sites on the water surface, (b) and (d) plasmodium minimized tube lengths.

evidence of pushing and pulling of single floating objects by pseudopodia. These are exemplified in Figs. 12.6 and 12.7.

To demonstrate pushing, we placed a piece of plastic foam on the water surface near the plasmodium (Fig. 12.6a). The plasmodium develops a pseudopodium which propagates towards the piece of foam (Fig. 12.6b). Due to the gravity force acting on the pseudopodium, a ripple is formed on the water surface (Fig. 12.6c), which pushes the piece of foam away from the growing pseudopodium's tip (Fig. 12.6d). There are no nutrients on the pushed object; therefore, the plasmodium abandons its attempt to occupy this piece of foam and retracts the pseudopodium (Fig. 12.6e). The piece remains stationary after that. This is an experimental demonstration of shifting.

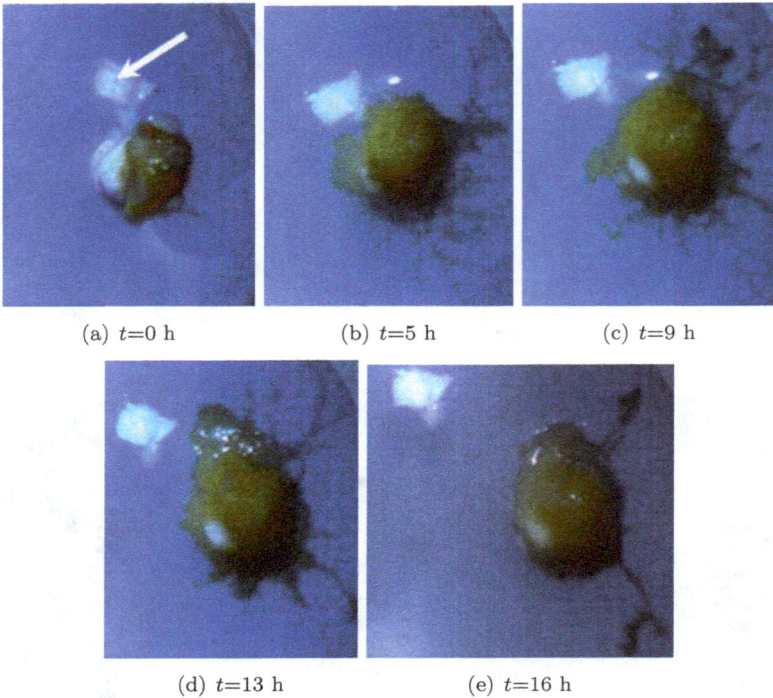

(a) $t=0$ h   (b) $t=5$ h   (c) $t=9$ h

(d) $t=13$ h   (e) $t=16$ h

Fig. 12.6   The plasmodium pushes lightweight floating objects. The pushed object is shown by white arrow.

Figure 12.7 illustrates pulling of a lightweight object. The piece of foam to be pulled is placed between two anchored objects (Fig. 12.7a). We inoculate plasmodium on one of these two stationary objects. The other stationary object receives an oat flake to attract the plasmodium.

A pseudopodium grows from its original location towards the site with the source of nutrients. The pseudopodium occupies the piece of foam (Fig. 12.7b) and then continues its propagation towards the source of nutrients (Fig. 12.7c). When the source of nutrients is reached (Fig. 12.7d), the protoplasmic tube connecting the two anchored objects contracts and straightens. The contraction causes the lightweight object to be pulled towards the source of nutrients (Fig. 12.7e).

Finding 21.   *Plasmodium growing on a water surface can push and pull floating objects. We can program plasmodium to manipulate floating objects by labeling the manipulated objects with chemo-attractants.*

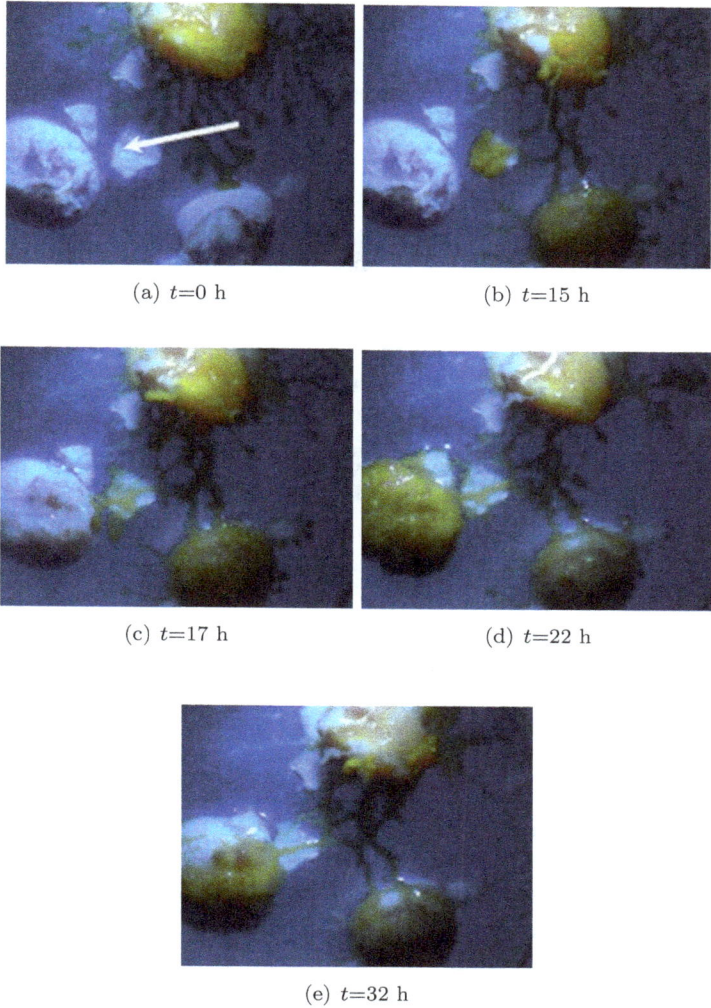

(a) $t=0$ h

(b) $t=15$ h

(c) $t=17$ h

(d) $t=22$ h

(e) $t=32$ h

Fig. 12.7   Plasmodium pulls lightweight object. The object to be pulled is indicated by white arrow.

The pushing and pulling capabilities of the plasmodium can be utilized in constructions of water-surface-based distributed manipulators [Hosokawa et al., 1996; Adamatzky et al., 2005].

12.3 Summary

Inspired by biomechanics of surface-walking insects, see e.g. [McAlister (1959); Suter et al. (1997); Suter (1999); Suter and Wildman (1999)], our previous studies of implementation of computing tasks in the plasmodium [Adamatzky (2007)]–[Adamatzky (2010)] and our ideas on design and fabrication of biological amorphous robots [Kennedy et al. (2001)], we explored the operational potential of plasmodium on a water surface. In experiments, we showed that the plasmodium inoculated on a water surface

- senses data objects represented by sources of nutrients,
- calculates the shortest path between the data objects on the water surface and approximates spanning trees of the data objects,
- pushes and pulls lightweight objects placed on the water surface.

Why a water surface? There are minimal friction and cohesion between the plasmodium's pseudopodia and the water surface. Also, the water surface provides good conditions for the plasmodium: always near-ideal humidity for the plasmodium, continuous removal of metabolites and excretions from the plasmodium's body and protoplasmic tubes.

Our experiments also show that a Physarum implementation of a Kolmogorov–Uspensky machine (Chap. 7.7) can be extended to a mechanical version of the storage modification machine by adding PUSH NODE and PULL NODE operations.

To translocate nodes selectively in the storage structures, we may need to assign certain attributes to the nodes. This can be done by marking nodes with different species of colors; in [Adamatzky (2008)] we demonstrated that the plasmodium exhibits strong preferences to some food colorings, and it is indifferent to others. Some food colorings repel the plasmodium. Such preference hierarchy can be mapped onto the mobile data storage structure. In the future, more experiments are indeed required to develop ideas derived from our scoping experiments to full working prototypes of the Physarum robots and mechanical Kolmogorov–Uspensky machines.

Chapter 13

Physarum boats

There were just two attempts to integrate spatially extended nonlinear chemical or biological systems with silicon hardware robots. In 2003, a wheeled robot controlled by an on-board excitable chemical system, a Belousov–Zhabotinsky medium, was successfully tested in the experimental arena [Adamatzky et al. (2004)]. An on-board chemical solution was stimulated by a silver wire and a vector towards the source of stimulation was extracted by the robot from the topology of the excitation waves in the stimulated Belousov–Zhabotinsky medium. In 2006, Tsuda et al. [Tsuda et al. (2007)] designed and successfully tested a Physarum controller for a legged robot. The controller's functioning was based on the fact that light inhibits oscillations in illuminated parts of the plasmodium. Thus, a direction towards the light source can be calculated from phase differences between shaded and illuminated parts of the plasmodium controller. In both instances nonlinear media controllers were coupled with conventional hardware and relied upon additional silicon devices to convert spatio-temporal dynamics of excitation and oscillation to the robot's motion.

An adaptive wet robot with no 'hardware' components and distributed intelligence would be an elegant way forward. Currently, we are aware only of non-intelligent droplets and quite rambling plasmodium.

Several experimental versions of externally guided droplets were proposed in [Sumino et al. (2005); Diquet (2009); Nagai et al. (2005)]. Ideologically, they are similar to the original ideas of [Yokoi and Kakazy (1992)] on guiding, splitting and merging mercury droplets, on an array of electrodes, by varying the potential between the electrodes. An oil droplet travels along acid-treated areas of a glass surface [Sumino et al. (2005)]. The droplets can be split and forced to merge again [Nagai et al. (2005)], pretty much as liquid metal T-1001. In experiments discussed in [Sumino

et al. (2005)], the pre-arranged path for a droplet is stationary. A dynamical control of a droplet motion is suggested in [Diquet (2009)]: an oleic
acid droplet propagates along a programmable path on a surface of photosensitive surfactant [Diquet (2009)]. The surfactant changes between cis
and trans configurations depending on the light wavelength. So, the path
for the droplet is projected onto the surfactant by light [Diquet (2009)].

The only result known to date of self-propelled excitable medium devices
can be attributed to Kitahata [Kitahata (2006)], who in 2005 experimentally showed that a droplet of Belousov–Zhabotinsky medium is capable of
translational motion due to changes in interfacial tension caused by convection forces inside the droplet. The convection forces are induced by
excitation waves propagating inside the droplet [Kitahata et al. (2002)].

In Chap. 11.6, we provided experimental evidence that plasmodium
can act as a distributed manipulator when placed on a water surface with
lightweight objects scattered around. We showed that the plasmodium
senses data objects, calculates the shortest path between the objects and
pushes and pulls lightweight objects placed on a water surface. We also
found that the motility of the plasmodium placed *directly* on the water surface is restricted. The plasmodium does propagate pseudopodia and forms
protoplasmic trees,; however, it does not travel as a localized entity. To
achieve true mobility of a plasmodium, we decided to attach it to a floater.
A priori, we hoped that due to mechanical oscillations of propagating pseudopodia the plasmodium will be capable of applying enough propulsive force
to the floater to propel the floater on the water surface. In laboratory experiments, we proved feasibility of the approach in [Adamatzky (2010)]. In
the present chapter, we overview our findings on motion patterns observed
in experiments and simulations.

Experiments were undertaken in Petri dishes with base diameter 35 mm,
filled by 1/3–1/5 with distilled water. Variously sized but not exceeding 5–
7 mm in longest dimension pieces of plastic and foam were used as floaters
residing on the water surface. Oat flakes occupied by the plasmodium
were placed on top of the floaters. The behavior of the plasmodium–floater
systems was recorded using a QX-5 digital microscope, magnification ×10.

The Petri dish with plasmodium was illuminated using light-emitting
diodes (LEDs) by white light which acted as a negative stimulus for the
plasmodium photo-taxis. The illuminating LEDs were positioned 2–4 cm
above the north-eastern part of the experimental container. The microscope
was placed in a dark box to keep the illumination gradient in the dish safe
from disturbances.

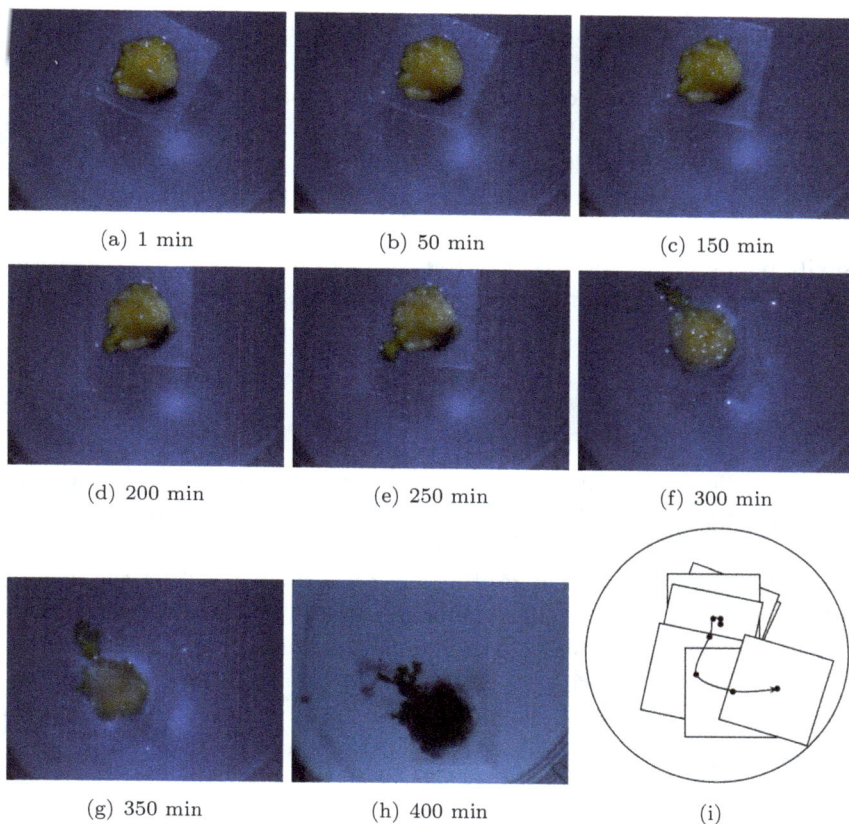

(a) 1 min   (b) 50 min   (c) 150 min

(d) 200 min   (e) 250 min   (f) 300 min

(g) 350 min   (h) 400 min   (i)

Fig. 13.1 Transition between types of motion of Physarum–floater system. Random wandering, or vibration-based, movement (a)–(e), 90° clockwise rotation (e) and (f) and directed propelling (g) and (h). Positions of the floater, at time steps corresponding to snapshots (a)–(h), and trajectory of the floater's center are shown in (i).

Finding 22. *A plasmodium–floater system executes the following types of movement: random wandering, quick sliding, pushing and directed propelling.*

13.1 Random wandering

The random wandering is caused by sudden movements, and associated vibrations, of plasmodium's protoplasm. Movement of protoplasm is caused by peristaltic contraction waves generated by a disordered network of bio-chemical oscillators. The higher the frequency of contractions in a partic-ular domain of the plasmodium, the more the protoplasm is attracted into the domain [Miyake et al. (1996); Nakagaki et al (1999)]. Relocations of protoplasmic mass change the mass center of the plasmodium–floater system, thus causing the floater to wander. The floater influenced by the protoplasmic vibration may exhibit a random motion (Fig. 13.1a–e), which usually persists until some part of the plasmodium propagates beyond the edges of the floater. After that, other types of motion take place.

13.2   Sliding

The quick sliding motion occurs when a substantial amount of protoplasm, e.g. a propagating pseudopodium, relocates to one edge of the floater and this pseudopodium penetrates the water surface. The floater then becomes partly submerged. This changes the way surface forces act on the floater, and sets the floater in motion. The floater usually slides along the surface until it collides with an obstacle or a wall of the container.

The random wandering and sliding are types of motion with non-controllable and unpredictable trajectories. These motions are unlikely to be used in future architectures of biological robots.

13.3   Pushing

The pushing motion can be observed in situations when plasmodium prop-agates from the floater onto the water surface and develops a tree of pro-toplasmic tubes. At some stage of the development, tips of the growing protoplasmic tree reach walls of the container and adhere to the walls. An example is shown in Fig. 13.2: 3–4 h after being placed on a floater the plasmodium starts propagation (Fig. 13.2b and c) and a protoplasmic tree emerges (Fig. 13.2f). Already at this stage the floater comes into motion (Fig. 13.2f and g), traveling in the direction opposite to propagation of the protoplasmic tree. The motion becomes pronounced when a part of the tree attaches itself to a wall of the container (Fig. 13.2h). A force of protoplasm pumped into the tubes causes the floater to noticeably move outwards from the growing tubes (Fig. 13.2i).

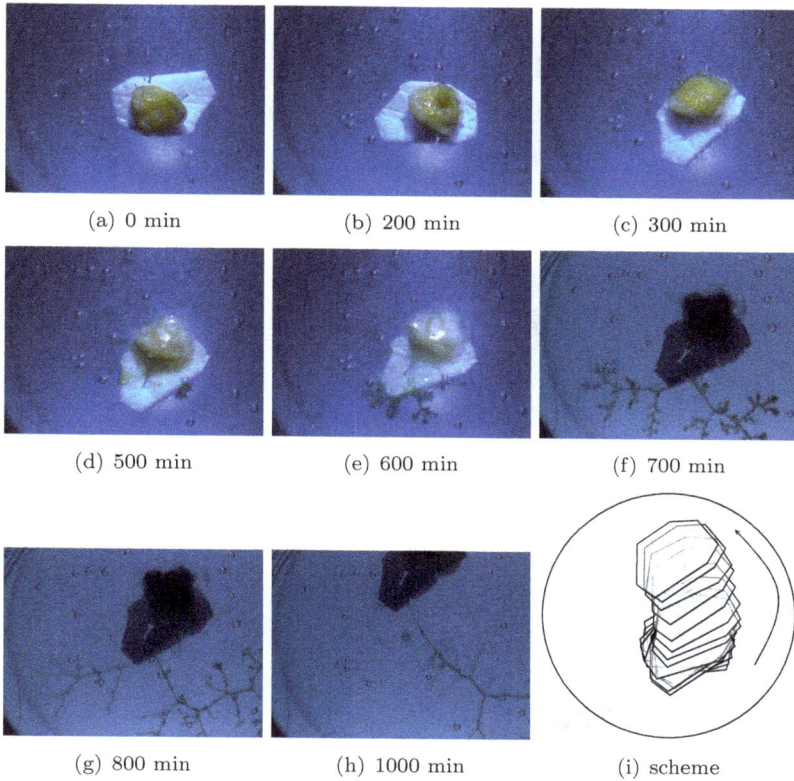

Fig. 13.2 Pushing the floater by growing protoplasmic tubes: (a)–(h) photographs of experimental container, (i) time-lapse contours of the floater.

13.4 Anchoring

A position of the floater can be 'fine tuned' by several protoplasmic trees, which attach themselves to different sides of the experimental container (Fig. 13.3). In the experiment shown in Fig. 13.3, protoplasmic trees grow along the north-western to south-eastern arc and towards the north-east (Fig. 13.3b and c). The sub-tree growing in the north-eastern direction diminishes with time, because it happens to be in the most illuminated area of the container. The only branch heading south survives. The protoplasmic tree becomes stronger in the north-western direction (Fig. 13.3b, c, g and h). The dynamical reallocations of the trees cause the floater to rotate and then gradually move south-west (Fig. 13.3i).

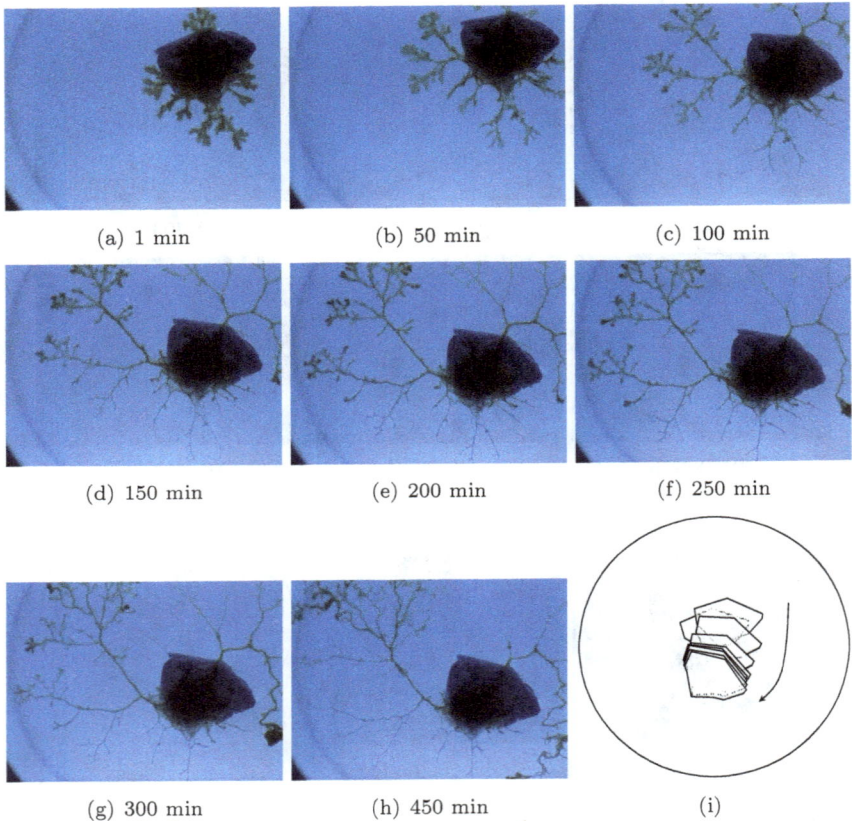

(a) 1 min   (b) 50 min   (c) 100 min

(d) 150 min   (e) 200 min   (f) 250 min

(g) 300 min   (h) 450 min   (i)

Fig. 13.3   Collective movement of the floater by several protoplasmic trees: (a)–(h) photographs of the experimental container, (i) time-lapse contours of the floater.

13.5   Propelling

The last and the most interesting type of motion observed is direct propelling of a floater by on-board plasmodium. This type of motion occurs when pseudopodia propagate beyond the edge of the floater but then partly sink. The pseudopodia oscillate. Their oscillations are clearly visible in the video recordings. These oscillations cause the floater to move in the direction opposite to the submerged pseudopodia. In contrast to sliding motion

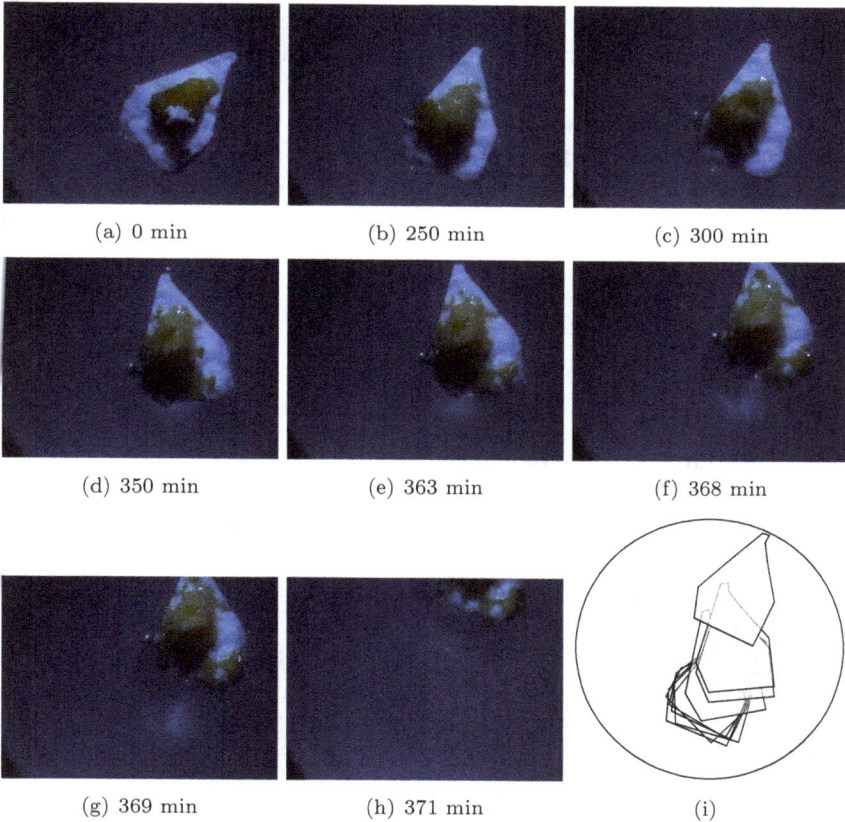

(a) 0 min   (b) 250 min   (c) 300 min

(d) 350 min   (e) 363 min   (f) 368 min

(g) 369 min   (h) 371 min   (i)

Fig. 13.4 Direct propelling of floater by oscillating pseudopodia: (a)–(h) photographs of the experimental container. (i) Time-lapse contours of the floater.

(Sect. 13.2), the floater does not submerge at all during the propulsive motion. Examples of direct propelling are shown in Figs. 13.1 and 13.4.

In the experiment shown in Fig. 13.1, plasmodium propagates a pseudopodium in the south-western direction (Fig. 13.1d and e). The pseudopodium does not stay on the water surface but partly sinks. The initial wetting of the pseudopodium causes the floater to rotate clockwise with the protruding pseudopodium facing now north-west (Fig. 13.1d and e). The immersed pseudopodia oscillate and propel the floater in the south-eastern direction (Fig. 13.1g–i).

An example of collective steering of the floater by several pseudopodia is shown in Fig. 13.4. After the initial adaptation period (Fig. 13.4a and b), associated with vibrational random motion, the plasmodium propagates pseudopodia in the western and south-eastern directions (Fig. 13.4d and e). At the beginning, only south-eastern pseudopodia contribute to propelling of the floater. Western pseudopodia are merely a balancing action of their south-eastern counterparts. Thus, the floater travels towards the north (Fig. 13.4f and g). Later, western pseudopodia increase the frequency of their oscillations and the floater turns towards the north-east (Fig. 13.4h and i).

Finding 23. *A floater propelled by plasmodium moves towards a source of illumination.*

13.6   Cellular automaton model

Plasmodium–floater systems are simulated by mobile two-dimensional cellular automata, with an eight-cell neighborhood, which produce force fields [Adamatzky and Melhuish (2002)]. We employ the model of retained excitation [Adamatzky et al. (2007)] as follows. Every cell of the automaton takes three states — resting, excited and refractory — and updates its state depending on states of its eight closest neighbors. A resting cell becomes excited if the number of its excited neighbors lies in the interval $[\theta_1, \theta_2]$, $1 \le \theta_1, \theta_2 \le 8$. An excited cell remains excited if the number of excited neighbors lies in the interval $[\delta_1, \delta_2]$, $1 \le \delta_1, \delta_2 \le 8$; otherwise, the cell takes the refractory state. A cell in the refractory state becomes resting in the next time step. This is a model of 'retained excitation' [Adamatzky et al. (2007)] used to imitate *P. polycephalum* foraging behavior [Adamatzky (2007a)]. This particular cellular automaton is good because it exhibits 'amoeba-shaped' patterns of excitation. We denote the local transition function as $R(\theta_1, \theta_2, \delta_1, \delta_2)$.

The reaction to a light stimulus is implemented as follows. Cells of lattice edges most distant from the light are excited with probability 0.15. The illumination-dependent excitation of edge cells corresponds to stimulus-dependent changes in plasmodium oscillation frequencies [Miyake et al. (1996); Nakagaki et al (1999)]. Namely, the parts of the plasmodium closest to positive stimulus (e.g. sources of nutrients) periodically contract with higher frequencies than parts closest to negative stimulus (e.g. high illumination).

We convert the automaton's excitation dynamics to motion by supplying every cell of the lattice with a virtual local force vector. The local vectors are updated at every step of the simulation. A vector in each cell becomes oriented towards a less excited part of the cell's neighborhood. An integral vector, which determines lattice rotation and translation at each step of simulation time, is calculated as a sum of local vectors over all cells. The approach was proved to be successful in simulation of mobile excitable lattices; see details in [Adamatzky and Melhuish (2002)].

Proposition 13.1. *Given a large enough container and a stationary source of light, a floater with plasmodium on board will move in irregular cycles around the source.*

By assuming that the container is large enough, we secured the situation when it is impossible for pseudopodia, or protoplasmic trees, to reach walls of the container and push the floater. The wandering motion only occurs at the very beginning of the plasmodium's development on the floater. Thus, we are left with propelling motion.

Let the floater be stationary. Due to negative photo-taxis, the plasmodium's pseudopodia propagate towards the less illuminated parts of the container. When the pseudopodia spawn from the floater to the water surface, their oscillatory contractions propel the floater towards the source of illumination. The pseudopodia continue their action until the floater passes the highly illuminated part. Then the pseudopodia happen to be on the most illuminated side of the floater. They retract and new pseudopodia are formed on the less illuminated part of the floater. They propel the floater towards the source of illumination again. The exact positions of growing pseudopodia depend on the distribution of biochemical sources of oscillations and interactions between traveling waves of contraction. New pseudopodia may not emerge at the same place as old pseudopodia. Thus, each trajectory of the floater towards the source of illumination will be different from previous ones.

To verify the proposition, we simulated a plasmodium–floater system by a mobile lattice with various cell-state transition functions. An illustrative series of snapshots is shown in Fig. 13.5. The automaton lattice is placed at some distance from the source of light (Fig. 13.5a). Edges of the lattice oriented towards the less illuminated area of the experimental container become excited (Fig. 13.5b). The excitation propels the lattice towards the source of light (Fig. 13.5e and f). When the lattice passes the site with highest illumination, another edge of the lattice excites and propels the lattice back to the domain of high illumination (Fig. 13.5g and h). The

(a) $t = 4$ (b) $t = 100$ (c) $t = 800$

(d) $t = 1400$ (e) $t = 2000$ (f) $t = 3800$

(g) $t = 6650$ (h) $t = 9000$

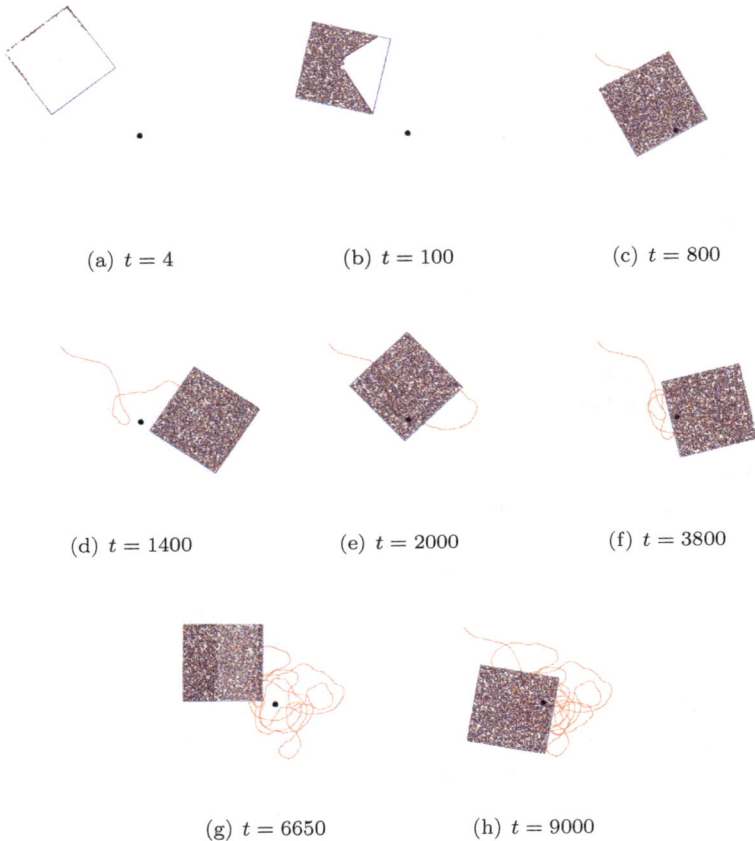

Fig. 13.5   Snapshots of cellular automaton model (two-dimensional lattice of 200×200 cells) of plasmodium–floater system. The local excitation dynamics are controlled by the function $R(2201)$. Source of illumination is shown by solid black disk. Each snapshot is supplied with trajectory of the center of the floater from the beginning of simulation.

sequence of excitations continues and the lattice travels along irregular cyclic trajectories around the source of illumination.

The exact pattern of lattice cycling around the source of illumination is determined by particulars of the local excitation dynamics. For example, trajectories of a lattice with threshold excitation (Fig. 13.6a) are very compactly arranged around the source. As soon as the lattice approaches the source of illumination it stays nearby, mostly turning back and forth. When we impose an upper boundary of the excitation, e.g. when a resting

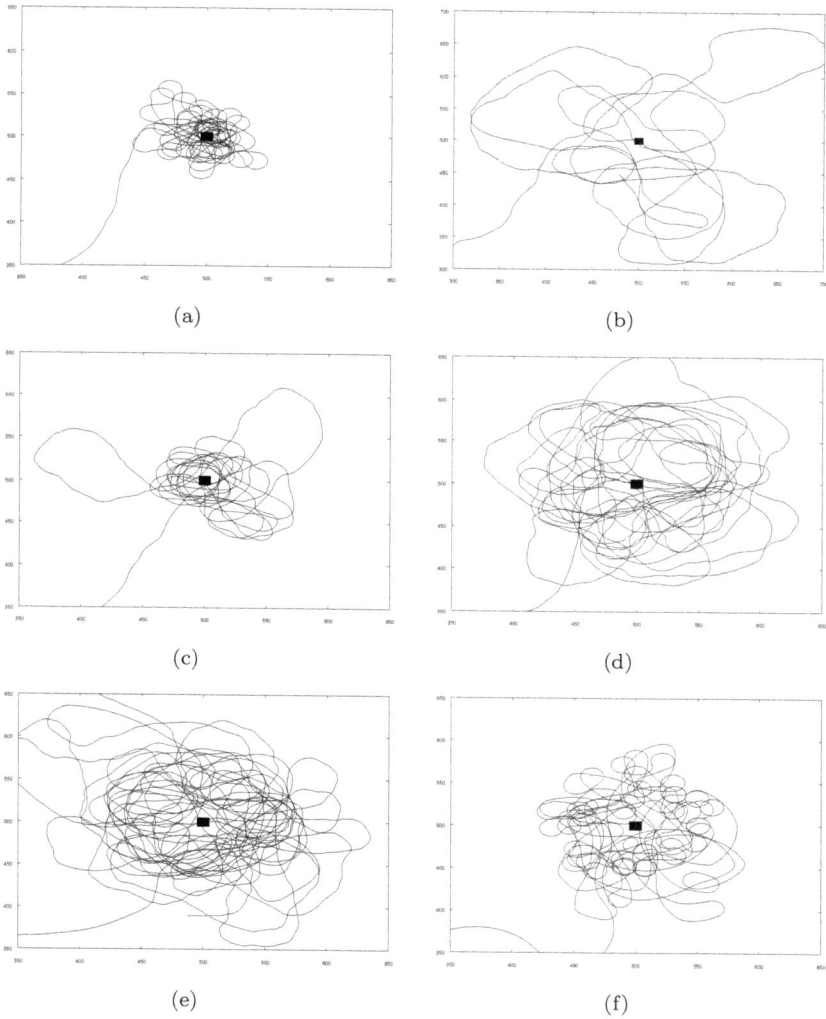

Fig. 13.6 Trajectories of the center of cellular automaton lattice traveling around source of illumination. The lattice's behavior is governed by local transition functions as follows: (a) $R(1899)$, threshold excitation, a resting cell excites if there is at least one excited neighbor, value 9 for δ_1 and δ_2 mean that the cell never stays excited longer than one step at a time, (b) $R(1299)$, a resting cell excites if it has one or two excited neighbors, (c) $R(2222)$, (d) $R(2201)$, (e) $R(2211)$, (f) $R(2246)$.

cell is excited only if one or two of its neighbors are excited, the lattice trajectory loosens (Fig. 13.6b).

Let us consider a lattice governed by the function $R(\theta_1\theta_2\delta_1\delta_2)$. A resting cell excites if the number σ of excited neighbors lies in the interval $[\theta_1, \theta_2]$ and the excited cell stays excited if $\sigma \in [\delta_1, \delta_2]$ (Fig. 13.6c–f). We focus on the excitation interval $[2, 2]$ because this type of local transition is typical for media with traveling localized excitations [Adamatzky (2001)], which is closely related to propagation activities of plasmodium of *P. polycephalum* [Adamatzky et al. (2008)]. We observed that lattices with very narrow intervals of excitation and retained excitation demonstrate a combination of compact and loose trajectories (Fig. 13.6c). Widening the interval of retained excitation disperses the lattice's trajectories in space (Fig. 13.6d and e). This emulates delayed responses to changes in sensorial background. When lower, δ_1, and upper, δ_2, boundaries of the retained excitation increase to 4 and 6, respectively, a lattice starts to exhibit a quasi-ordered behavior: a combination of long runs away from the source of illumination and tidy loops of rotations (Fig. 13.6f).

13.7   Physarum tugboat

A floater propelled by plasmodium exerts enough force to push other floating objects. Experimental evidence is provided in Fig. 13.7. Two pieces of plastic film are placed on the water surface. We place an oat flake on one piece and a virgin oat flake on the other (Fig. 13.7a). Trying to escape from the source of illumination, positioned north-east, the plasmodium develops a pseudopodium towards the west-south-west (Fig. 13.7b–d). Periodic waves of contraction running along the pseudopodium cause it to make beating motions, thus transferring a propulsive force to the floater. The floater moves east-north-east and collides with another floater (Fig. 13.7e). The Physarum 'tugboat' continues pushing another floater until it touches the wall of the Petri dish (Fig. 13.7f–i).

13.8   On failures

We do not want readers to fall into a trap of perfection. Experiments with biological objects are never ideal. Let us look at two examples of experiments which did not go as planned.

When trying to propagate towards a shady area, the plasmodium displaced too much weight outside the floater (Fig. 13.8a and b). The pseudopodia and part of the floater thus partially submerge under the wa-

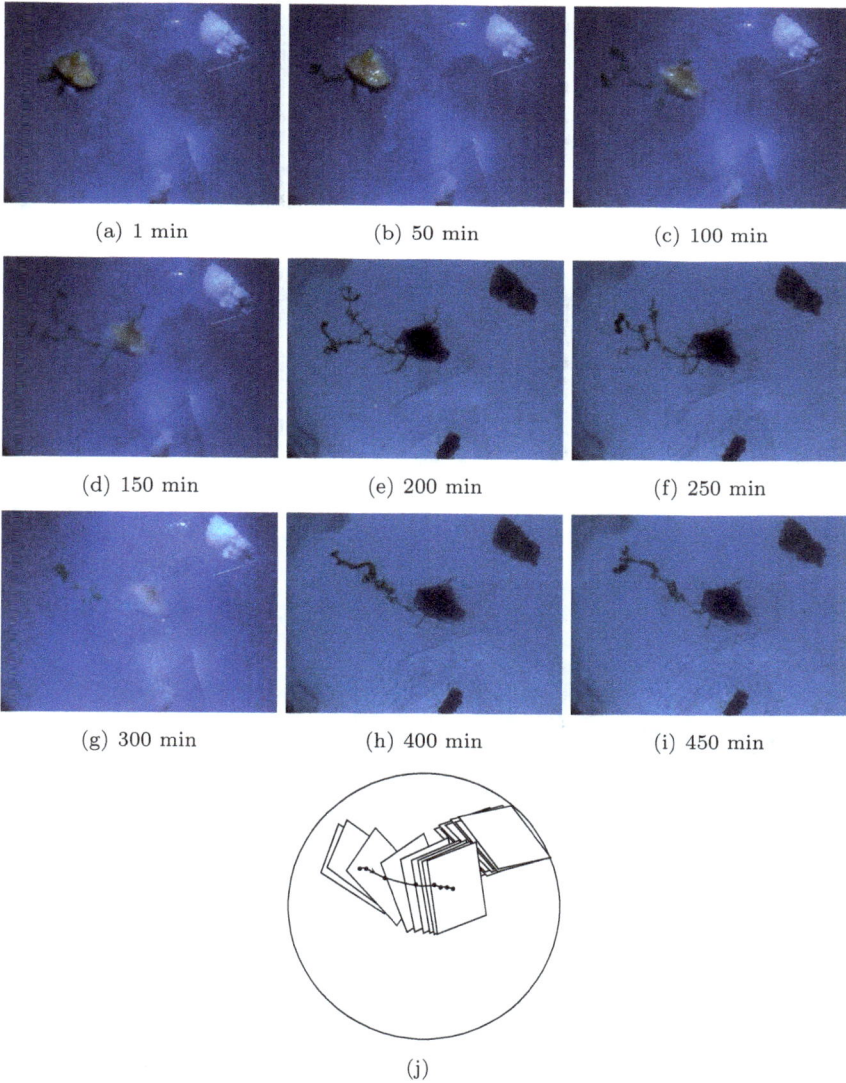

(a) 1 min

(b) 50 min

(c) 100 min

(d) 150 min

(e) 200 min

(f) 250 min

(g) 300 min

(h) 400 min

(i) 450 min

(j)

Fig. 13.7   (a)–(i) Snapshots of Physarum tugboat, (j) time-lapse contours of the floaters. (a)–(d), (g) images of experimental Petri dish illuminated from above; (e), (f), (h), (i) images of experimental Petri dish illuminated from bottom.

ter surface (Fig. 13.8c and d). Due to a combination of irregularities on the water surface, forces caused by the submerged floater and beating of

(a) 0 min   (b) 300 min   (c) 350 min

(d) 400 min   (e) 450 min   (f) 500 min

(g) 550 min   (h) 600 min   (i)

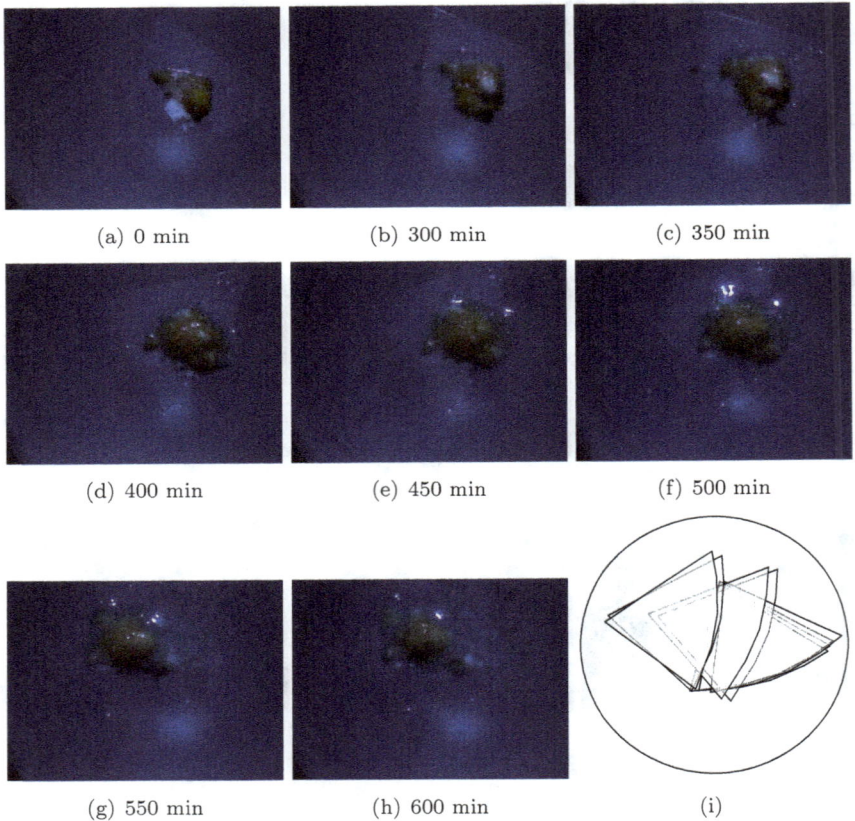

Fig. 13.8   Moving by submerging: (a)–(h) snapshots of experimental arena, (i) time-lapse contours of the floaters.

the pseudopodia the floater rotates clockwise (Fig. 13.8i) and moves west (Fig. 13.8e–i).

An anomalous scenario develops in Fig. 13.9. Plasmodium grows, on the water surface, in the direction of increased illumination (Fig. 13.9a–c). Its pseudopodium attaches itself to the Petri dish wall (Fig. 13.9d). Due to protoplasmic pressure in this tube, the floater rotates clockwise and moves away from the source of illumination (Fig. 13.9e–h).

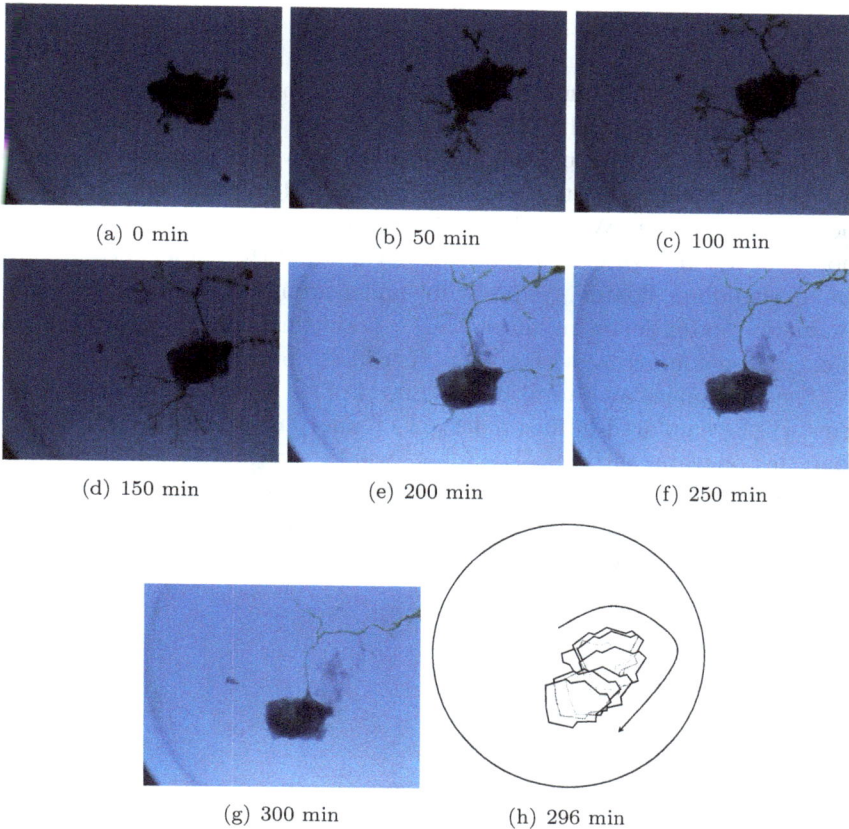

(a) 0 min   (b) 50 min   (c) 100 min

(d) 150 min   (e) 200 min   (f) 250 min

(g) 300 min   (h) 296 min

Fig. 13.9   Moving away from light: (a)–(g) snapshots of experimental arena, (h) time-lapse contours of the floater. All snapshots are made with bottom illumination to highlight morphology of pseudopodia.

13.9   Summary

In laboratory experiments and computer simulations we showed that plasmodium of *P. polycephalum* acts as an 'engine' or a propelling agent for lightweight floating objects. We found that the most typical type of movement generated by the plasmodium is a propulsive forward motion. This motion is caused by pseudopodia protruding beyond the the floater and oscillating. In a small container, when the growing tree of the plasmodium's protoplasmic tubes can reach the sides of the container, the plasmodium pushes the floater by increasing the length of the tubes connecting the floater with the container's sides.

A plasmodium shows negative photo-taxis. It tries to evade regions illuminated by non-yellow light and grows towards more shaded areas. When the plasmodium is attached to a floating object, the plasmodium–floater system exhibits positive photo-taxis. Due to growth, and associated oscillations, of pseudopodia on the less illuminated side of a floater, the floater moves towards light. We mimicked this phenomenon in cellular automaton models of mobile light-sensitive lattices. We found that in ideal conditions the 'plasmodium–floater' system will wander around the site of highest illumination, often following quasi-chaotic trajectories due to many sources of excitation competing with one another in the protoplasm.

The phenomena and primitive constructs of the plasmodium–floater systems will be employed in future designs of amorphous decentralized robots operating on water surfaces. The robots will be capable of searching for objects, traveling towards the objects' locations, implementing manipulation and sorting. The robots will be controlled by gradient fields of illumination and chemo-attractants.

Chapter 14

Manipulating substances with Physarum machine

Plasmodium of *P. polycephalum* constructs a range of proximity graphs — spanning trees, relative neighborhood graphs, Gabriel graphs and Delone diagrams (Chap. 5.5) — and approximates Voronoi diagrams (Chap. 3.3). Structures developed by plasmodium can form a basis for an intelligent adaptive network of distributing, mixing and manipulating of chemical substances, which are relatively harmless for plasmodium. Such an intelligent network may act as a 'lab-in-a-plasmodium' for experiments in biological micro-fluidics, drug testing devices [Terayama et al. (1978)], biosensors and distributed amorphous delivery system. In the present chapter, we apply our ideas of plasmodium-based amorphous robotics to evaluate manipulation of substances by the plasmodium's protoplasmic network.

Due to cytoplasmic streaming [Allen et al. (1963); Stewart and Stewart (1959); Hulsmann and Wohlfarth-Bottermann (1978); Newton et al. (1977); Gawlitta et al. (1980); Bykov et al. (2009)], a relatively harmless colored substance can be naturally ingested by plasmodium and distributed inside the protoplasmic network. By controlling the plasmodium's propagation over an uncolored substrate, we can 'fill' specified areas of the substrate with the color transported by the plasmodium.

14.1 Operations with colored substances

Let us develop a procedure for controllable transfer of colors between spatial domains and mixing colors by plasmodium. Let X, Y and L be finite compact domains of a Euclidean plane. Given X and Y domains filled with colors c_X and c_Y and domain L being uncolored, we want to color the domain L either with color c_X or c_Y or a compound color c mixed of c_X and c_Y.

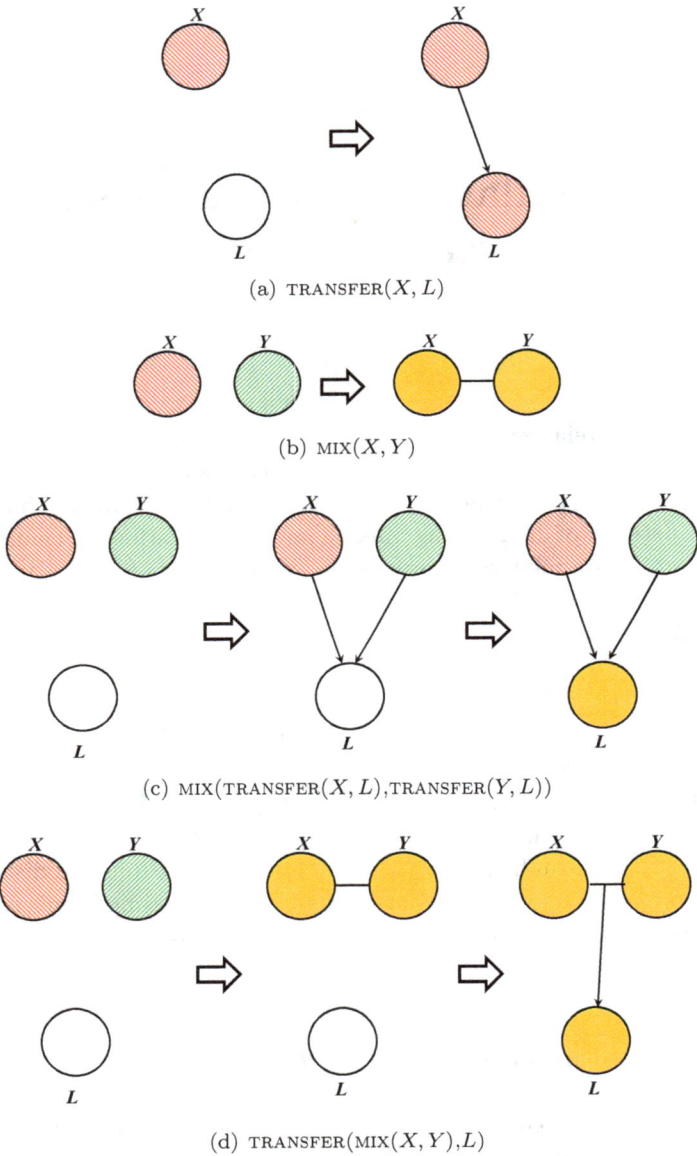

(a) TRANSFER(X, L)

(b) MIX(X, Y)

(c) MIX(TRANSFER(X, L),TRANSFER(Y, L))

(d) TRANSFER(MIX(X, Y),L)

Fig. 14.1 Basic operations to be realized by plasmodium of *P. polycephalum*: plasmodia are placed in the colored domains X and Y, domain L is initially uncolored.

To demonstrate that a color can be mixed and transferred in a controllable manner (pioneering experimental results on mixing colored substances were presented in [Nakagaki et al. (2000)]), we should experimentally implement the following basic operations (Fig. 14.1):

TRANSFER(X, L): plasmodium is inoculated in domain X and transports color c_X to domain L (Fig. 14.1a),

MIX(X, Y): plasmodia are inoculated in colored domains X and Y, the plasmodia merge and mix colors c_X and c_Y (Fig. 14.1b)

and their superpositions:

MIX(TRANSFER(X, L),TRANSFER(Y, L)): plasmodia are inoculated in colored domains X and Y, each plasmodium is independently guided towards domain L, the plasmodia merge with each other in the domain L and colors they brought into the domain become mixed (Fig. 14.1c, see experimental results in Sect. 14.4),

TRANSFER(MIX(X, Y),L): plasmodia are inoculated in colored domains X and Y, the plasmodia are encouraged to merge their protoplasmic networks and to mix their colors, the joint plasmodium is then guided towards domain L (Fig. 14.1d, see experimental results in Sect. 14.4).

The non-nutrient agar is used to 'program' localized transfer of substances. The plasmodium's behavior strongly depends on the presence/absence of nutrients in the substrate [Adamatzky et al. (2008)]. On a nutrient-rich substrate, e.g. corn meal agar, plasmodium grows in all directions, expanding circularly. On a non-nutrient substrate, e.g. agar or a humid filter paper, plasmodium propagates only in certain directions, mostly towards sources of nutrients.

Oat flakes are saturated with SuperCook Food Colorings [1]:

- green (tartrazine E102, green S E142),
- yellow (tartrazine E102, sunset yellow E110, Ponceau 4R E124),
- blue (brilliant blue E133, azorubin E122),
- red (sunset yellow E110, azorubin E122).

The flakes are quickly dried after saturation (Fig. 14.2). Depending on the particular experiment, a few colored oat flakes, and one uncolored oat flake colonized by the plasmodium, are placed on agar gel. To prevent diffusion of colorings from colored flakes to agar, closed cuts in the agar

---

[1] www.supercook.co.uk

Fig. 14.2   Dyeing oat flakes with food colorings. Photograph shows bottles with food colorings, and small heaps of oat flakes soaked with the colorings and then left to dry.

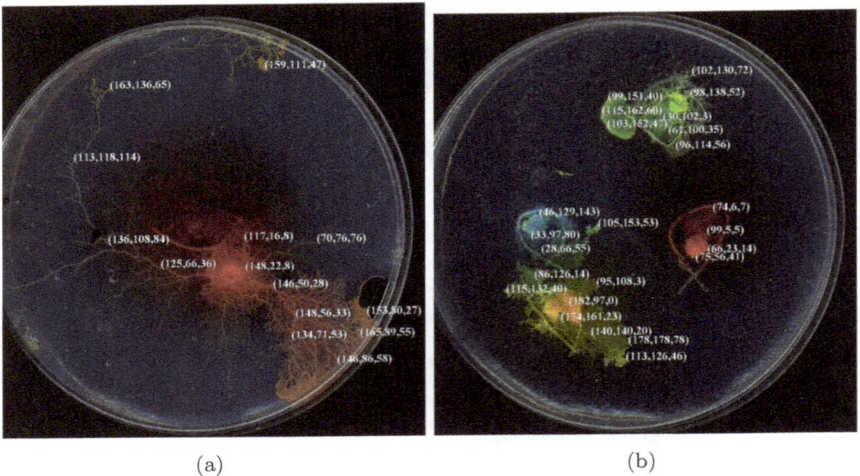

(a)   (b)

Fig. 14.3   Example distribution of coloring in plasmodium's protoplasmic tubes.

immediately surrounding the flakes are made. Plasmodium moves through the cuts in agar without problems while diffusion of colorings is restricted. In experiments on controlling plasmodium propagation with repelling fields, we use grains of coarse sea salt[2].

The Petri dishes are kept in the dark, at temperatures of 22–25°C, except for observation and image recording. Periodically, the dishes are

---

[2]Saxa coarse grain sea salt, RHM Foods, Middlewich, CW10 0HD, UK.

scanned using an Epson Perfection 4490 scanner. Raw[3] images of the plasmodia are analyzed using basic software coded in Processing[4]. The distribution of colorings in the plasmodium's body, pseudopodia and protoplasmic tubes is verified by extracting values of any particular pixel in additive color, RGB, mode. An example is provided in Fig. 14.3. Average natural RGB values of plasmodium not colored artificially, as measured in our experiments, are $(r, g, b) = (120, 130, 50)$, with usual dispersion of values $0 \leq (g - r) \leq 20$ and $60 \leq b \leq 80$.

In general, the intensity of the principal component of coloring in the plasmodium tubes is preserved sufficiently well. Let us consider an experiment of cultivating plasmodium on red-colored oat flakes (Fig. 14.4a). The distribution of RGB values of pixels along the longest protoplasmic tube (Fig. 14.4) shows that, despite some irregularities, the red component of the tube's color remains 1.6 times higher than the green and blue components (Fig. 14.4b). The green component starts to increase at a distance of more than 14 cm from the original colored oat flake. This indicates that eventually pigments inside the plasmodium undergo transformations and the plasmodium tends to return to its normal color in the long term.

In some experiments [Adamatzky (2007)], we observed that, when given a choice, the plasmodium prefers uncolored flakes to green-, yellow- and blue-colored ones. Colonization of red-colored flakes is the last option for the plasmodium. Such preferences are unstable and the exact choice made by the plasmodium depends on the distance to any particular flake, state of the flake's colonization by bacteria and local properties of the substrate. A few examples of plasmodium's choices of colored flakes are shown in Fig. 14.5.

In a first example, an oat flake colonized by plasmodium is placed in the center of a Petri dish (Fig. 14.5a). Yellow- and green-colored flakes are positioned nearby. The plasmodium first occupies a green-colored flake and only then propagates towards a yellow-colored flake. When propagating to the green flake, the plasmodium tries to avoid the yellow flake, which is reflected in the shape of the plasmodium's protoplasmic tube (Fig. 14.5a).

A second example, Fig. 14.5b, shows that in certain circumstances the plasmodium chooses the closest oat flake irrespectively of the flake's color. When the plasmodium is in a good physiological state it can attempt to colonize all flakes at once, as shown in Fig. 14.5b.

---

[3]To increase visibility of images for printing, but not for for analysis, the color saturation was increased.
[4]www.processing.org

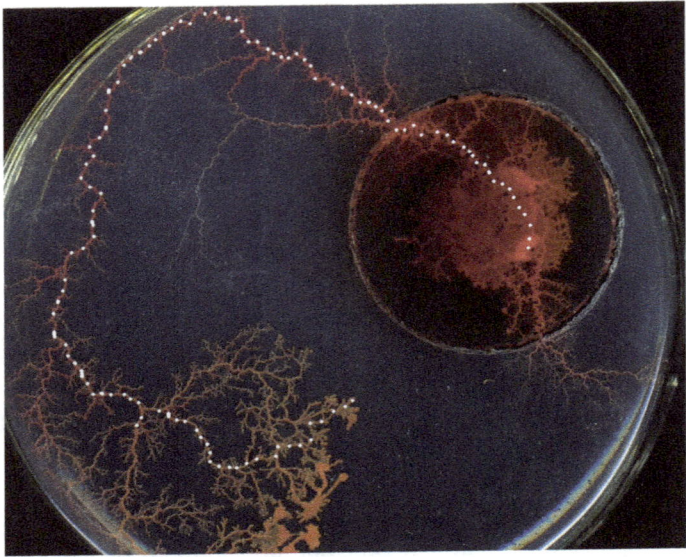

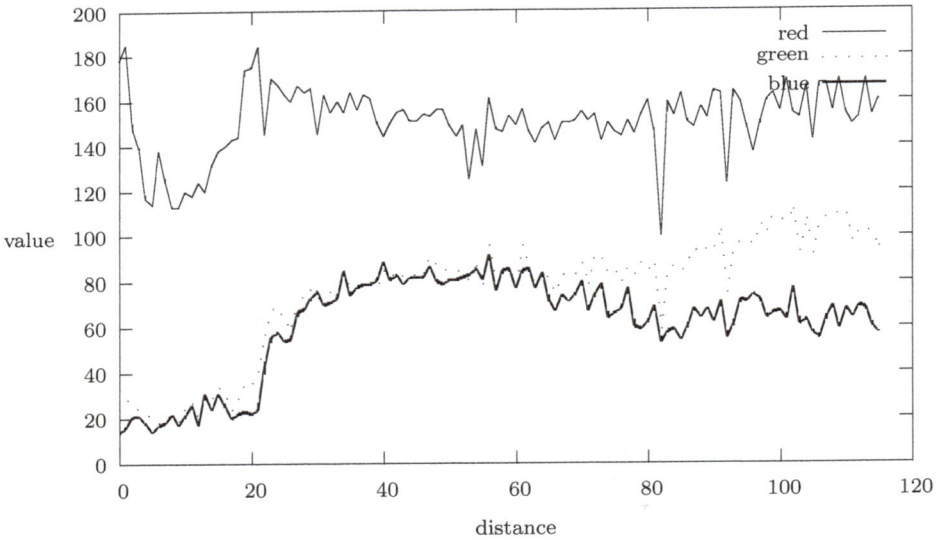

Fig. 14.4   Distribution of color values along protoplasmic tube. Distance between first measurement point and last measurement point *along* the protoplasmic tube is approx. 28 cm. Colored oat flakes, where plasmodium was inoculated initially, are in the gel disk in the north-eastern quadrant of the image.

(a)　　　　　　　　　(b)

(c)

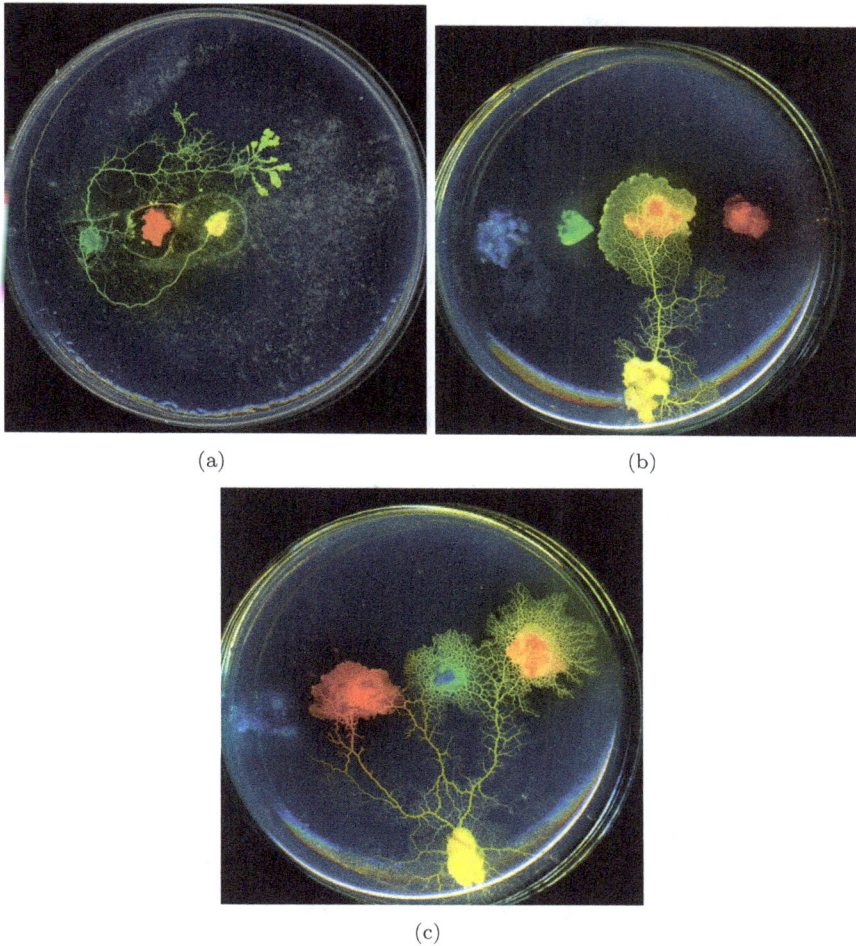

Fig. 14.5 Plasmodium starts to colonize colored flakes. The flakes in a row are colored as follows, from left to right: (a) green, yellow, uncolored, (b) blue, green, yellow, red, (c) blue, red, blue, yellow.

14.2 Transfer of substances to specified location

Let coloring C placed in X and L be an uncolored locus of the substrate, $X \cap L = \emptyset$. We wish to color L with C, i.e. transfer some quantity of C from domain X to domain L (Fig. 14.1a).

This can be realized in two steps. In the first step, inoculate the plas-

(a) $t = 0$ h   (b) $t = 12$ h

(c) $t = 28$ h

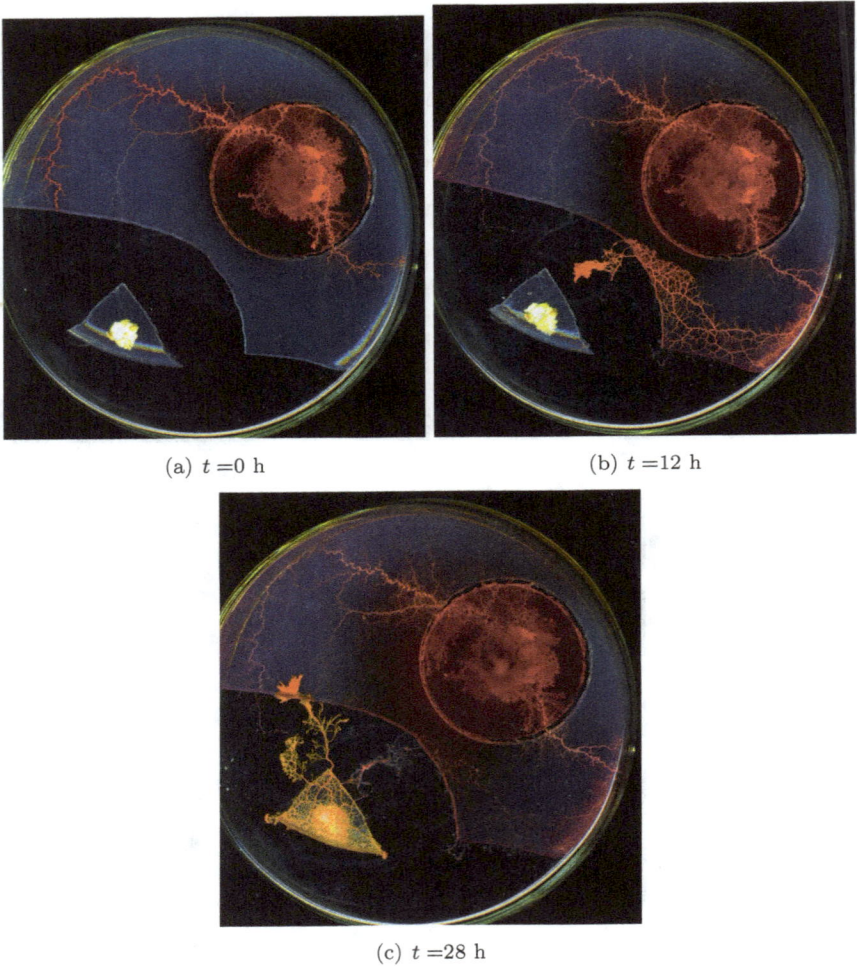

Fig. 14.6   Example of coloring using attracting stimulus.

modium in X, so it starts taking in coloring C. While consuming C, the plasmodium distributes C in its extended body. The second step would be to guide the plasmodium towards L, so it occupies L and thus L becomes colored with C. The plasmodium can be guided by attracting (Chap. 8.4) and repelling (Chap. 10.10) fields.

An attracting field can be implemented with oat flakes. We place a flake in the domain X (Fig. 14.6a). In a few hours plasmodium starts grcwing towards L (Fig. 14.6b). The plasmodium even propagates over an unfriendly dry substrate — the bottom of the plastic dish not covered by agar gel. Eventually, the plasmodium reaches L and 'paints' the domain with color C (Fig. 14.6c).

We can also 'fill' locus L with colored plasmodium without placing any sources of nutrients in L but by arranging a configuration of repelling sources to guide the plasmodium towards L. A repelling field can be implemented by applying a gradient of illumination to the substrate. Due to its light avoidance [Nakagaki et al (1999)], the plasmodium tends to propagate into less illuminated regions of the substrate. Light-controlled propagation of plasmodium is a good non-invasive technique of shaping a protoplasmic network (Chap. 9.4); however, the plasmodium does not always react predictably to an applied gradient of illumination. If we want a more 'pronounced' reaction to control stimuli, we can employ a more drastic way of controlling the plasmodium — with grains of sodium chloride. The first repelling field, or configuration of repellents, determines the initial vector of plasmodium propagation. Further positions of repellents act as reflectors to shape the trajectory of the plasmodium movement.

In the example shown in in Fig. 14.7, the wall of the Petri dish acts as a passive (i.e. without chemo-repellents) reflector. We place several grains of salt to direct the plasmodium's motion towards the westward proximity of the dish (Fig. 14.7a). The plasmodium collides with the western wall of the dish (Fig. 14.7b). We orient the plasmodium northward by placing additional grains of salt southward of the plasmodium (Fig. 14.7c). Fine tuning of the plasmodium's shape and propagation is achieved by placing two additional grains of salt (Fig. 14.7c and d).

By carefully arranging grains of salt, we can manipulate several propagating parts of a plasmodium at once. Thus, in Fig. 14.8, we provide experimental evidence of guiding a green-colored part of plasmodium westward (and then southward) and blue-colored plasmodium southward.

14.3   Mixing substances

If we place two plasmodia in a Petri dish, their protoplasmic tubes eventually merge. Thus, these two plasmodia will form a single cell. What happens when two plasmodia are cultivated on differently colored, e.g. $C_1 = (R_1, G_1, B_1)$ and $C_2 = (R_2, G_2, B_2)$, oat flakes and then merge? Will the newly formed single organism have an 'average' color $(R_1 + R_2, G_1 + G_2, B_1 + B_2)/2$ (Fig. 14.1b)?

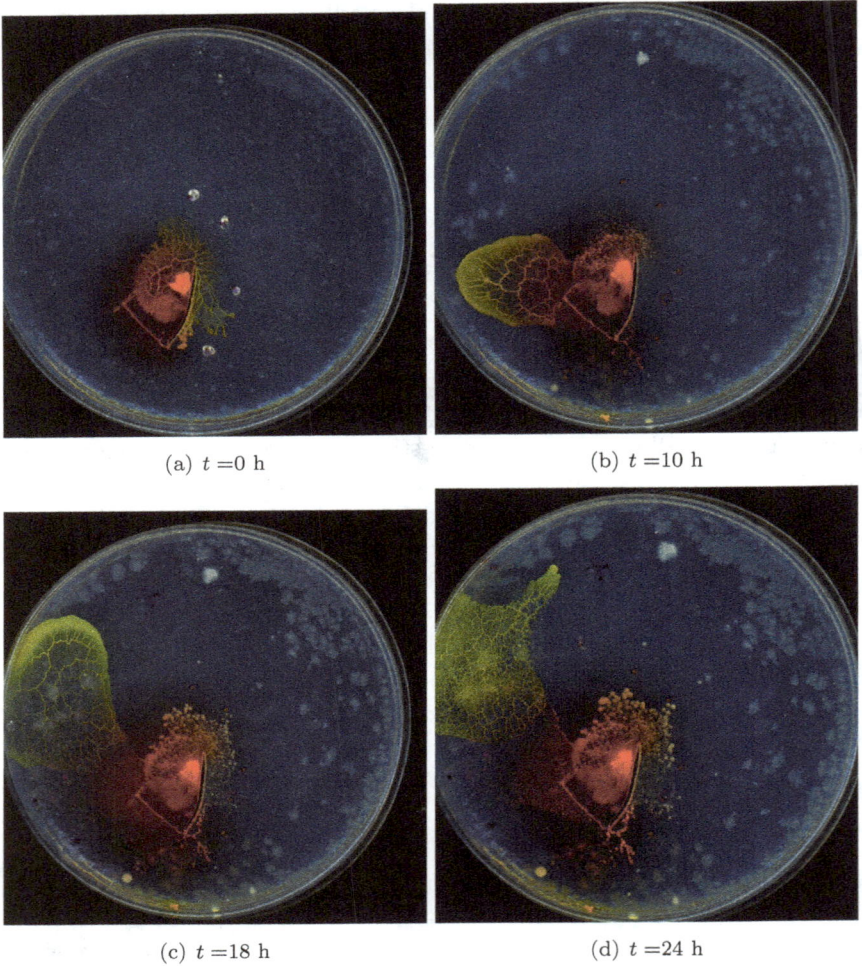

(a) $t =0$ h   (b) $t =10$ h

(c) $t =18$ h   (d) $t =24$ h

Fig. 14.7 Guiding plasmodium by repelling fields: (a)–(d) images of the experimental arena, (e) scheme of the plasmodium propagation: target is shown by 'sun' sign, trajectory of plasmodium propagation by dotted line, grains of salt by empty stars (added at 0 h), gray stars (10 h), black stars (18 h) and black disks (24 h).

We found that colorings are not always mixed in the short term but any particular tube usually takes the color of the closest colored flake. An example is provided in Fig. 14.9.

Initially, plasmodia are placed in red-colored (Fig. 14.9a on the left)

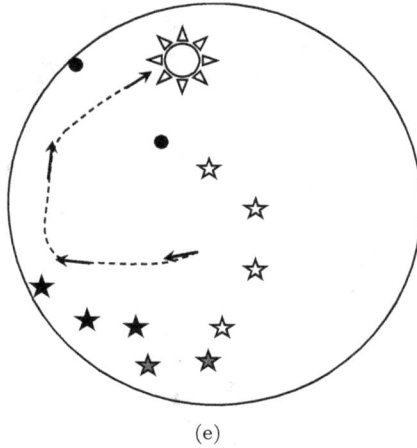

(e)

Fig. 14.7 *Continued.*

and blue-colored (Fig. 14.9a on the right) oat flakes. The plasmodia take in colorings from their unique flakes and grow towards each other. Eventually, their protoplasmic tubes merge. The distribution of color values along the thickest tube (Fig. 14.9b) shows that the value of the red component of RGB color increases and the value of the blue component decreases along the tube connecting the red-colored flake to the blue-colored flake.

Mixing color in a T-shaped gel plate is shown in Fig. 14.10. Plasmodia are placed in proximity of red- and yellow-colored oat flakes in western and eastern shoulders of a T-shaped gel plate. The plasmodia propagate, merge and spread towards the 'output' vertical part of the T-junction. Figure 14.10b shows an image with separated colors. Let p be a pixel of the original image and (r, g, b) the RGB values of the pixel p's color. The color of pixel p is selected by the following procedure:

(1) if $r > 150$ and $g > 120$, we assign p pure yellow color;
(2) if $(r - g) > 10$ and $r > 130$, p takes pure red color;
(3) if $r < 100$ and $g < 100$, p takes pure green color.

These conditions are checked in the priority (1) to (3), and if none of the conditions holds the pixel p is assigned white color. In the resultant image (Fig. 14.10b), we see that red and yellow colors are almost unmixed (apart from one large tube between red and yellow flakes) in the areas closest to the positions of initial inoculations. Red and yellow color separations are also shown in Fig. 14.10c and d. In Fig. 14.10e, parts of the plasmodium colored

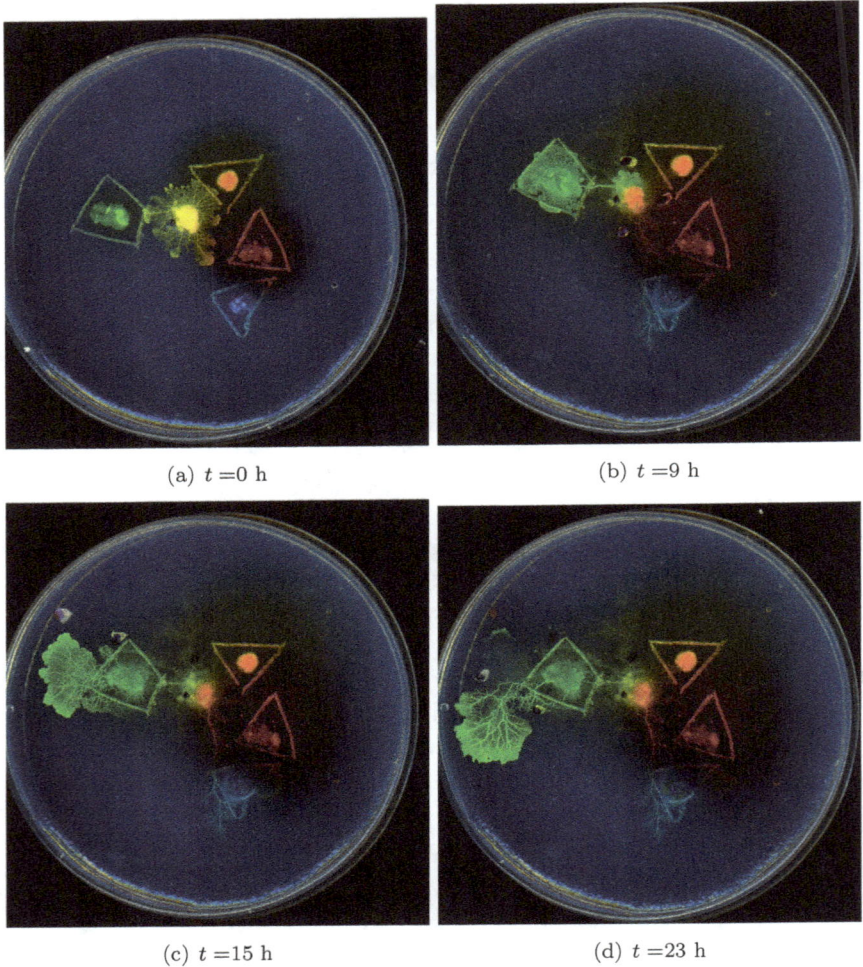

(a) $t = 0$ h

(b) $t = 9$ h

(c) $t = 15$ h

(d) $t = 23$ h

Fig. 14.8 Manipulating several plasmodium localizations at once: (a)–(e) scanned images of experiments, (f) scheme of the manipulation: locations of salt grains are shown by stars, trajectories of plasmodia movement by dotted lines.

in orange are shown, namely those pixels which satisfy the conditions $r > 170$ and $(r - g) > 60$.

The images shown in Fig. 14.11 demonstrate that during the protoplasmic network development the color of the closest colored flake becomes dominating.

Figure 14.12 provides an excellent example of how plasmodium mixes

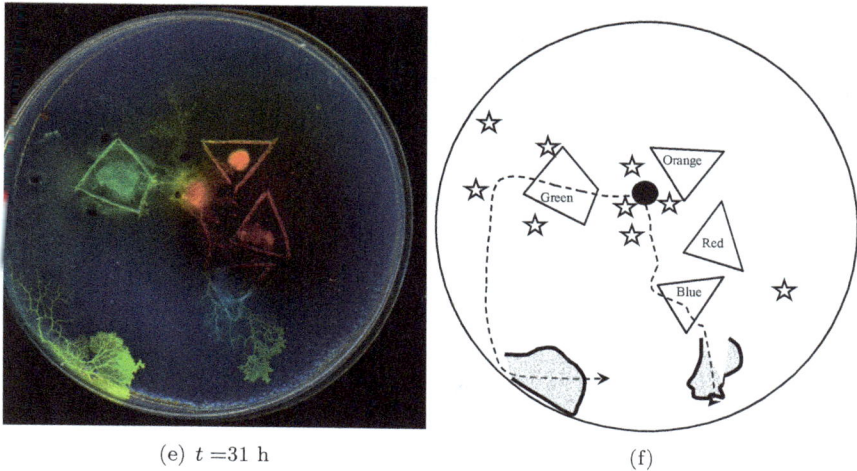

(e) $t = 31$ h   (f)

Fig. 14.8 *Continued.*

two colors to produce a third color. Red and green oat flakes are placed in the north-western and north-eastern parts of a Petri dish. The flakes are inoculated with plasmodia. The plasmodia propagate and merge their protoplasmic tubes (Fig. 14.12a and b). Major tubes connecting green and red domains show a transition from green to yellow to red colors (Fig. 14.12b–e). Also, a substantial localized part of the plasmodium starts propagating southward along the western part of the Petri dish. This traveling localization is colored in yellow; the average RGB values of pixel colors are $(170, 140, 40)$. The color produced differs from the natural yellow color of the plasmodium because it has a higher value of the red component and a lower value of the blue component than the natural color of the plasmodium.

14.4   Superpositions of TRANSFER and MIX operations

To paint any specified domain of the substrate with mixed color, not available in the original colored oat flakes, we can

- either force plasmodium to mix colors directly in the specified domain, operation MIX(TRANSFER(X, L),TRANSFER(Y, L)) (Fig. 14.1c)), or
- allow plasmodium to mix colors somewhere else and then guide the plasmodium carrying mixed colorings to the specified domain, operation TRANSFER(MIX(X, Y),L) (Fig. 14.1d).

(a)

(b)

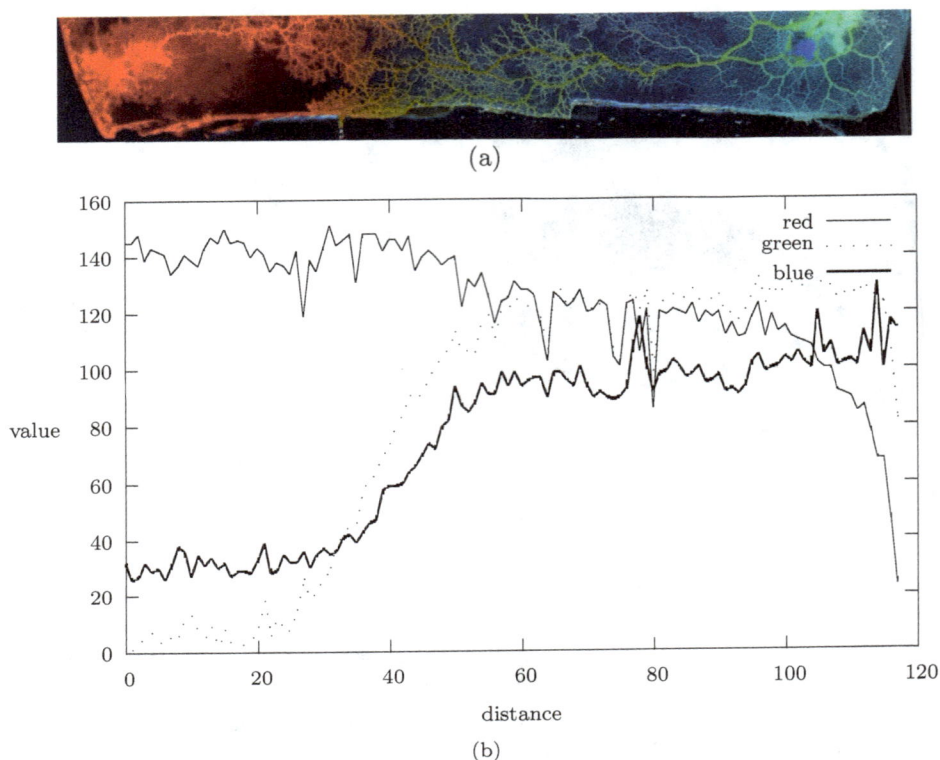

Fig. 14.9 Mixing of red and blue colorings in protoplasmic tubes of *P. polycephalum*: (a) image of the experimental mixing: red-colored oat flake is on the left, blue-colored oat flake is on the right, (b) distribution of RGB values of sample pixels along largest protoplasmic tube connecting red and blue oat flakes.

An example of forcing the plasmodium to mix colors directly in the specified domain is shown in Fig. 14.13. Assume that we have only red and green colorings, and we want to color the south-western part of the Petri dish with yellow color. We place plasmodia near red- and green-colored oat flakes. The plasmodia consume the colorings. When the plasmodia start propagating outwards from their colored flakes, we place grains of salt north-west of the red-colored plasmodium and north-east-east of the green-colored plasmodium. The plasmodia migrate to avoid a high concentration of sodium chloride. They collide with each other, merge and mix their colorings. The distribution of colors can be verified in Fig. 14.13c–e.

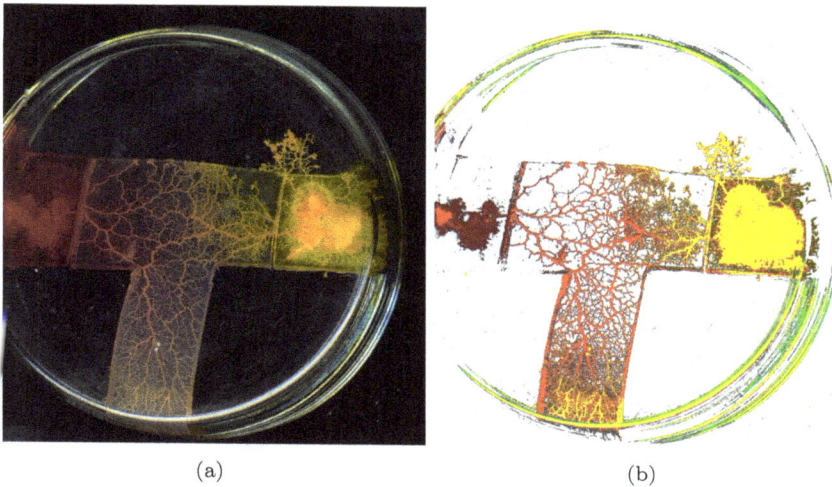

(a)   (b)

Fig. 14.10   Mixing colors by plasmodium: (a) scanned image of plasmodium, (b) colors are separated, (c) red-colored tubes, (d) yellow-colored tubes, (e) orange-colored tubes.

By increasing the concentration of repellents, we can make the plasmodium with mixed colors abandon all connections with its initial location and migrate as an autonomous blob. An example is shown in Fig. 14.14. Separate plasmodia are placed near red-colored and green-colored oat flakes. The plasmodia take in colorings and start propagating. At this moment, we place salt grains in northern, eastern and southern parts of the Petri dish (Fig. 14.14a and c). Red-colored and green-colored plasmodia propagate westward. They eventually merge. Due to the salt grains placed in the immediate proximity of the colored oat flakes, the plasmodia abandon tubes connecting them with their initial positions (Fig. 14.14b and d). Thus, we obtain yellow-colored autonomous plasmodium (Fig. 14.14b and d).

14.5   Summary

In laboratory experiments we demonstrated that plasmodium of *P. polycephalum* consumes food colorings and distributes them in its protoplasmic network. By specifically arranging the configuration of attractive — sources of nutrients — and repelling — increased concentration of sodium chloride — fields, we programmed the plasmodium to implement the following operations:

(c)   (d)

(e)

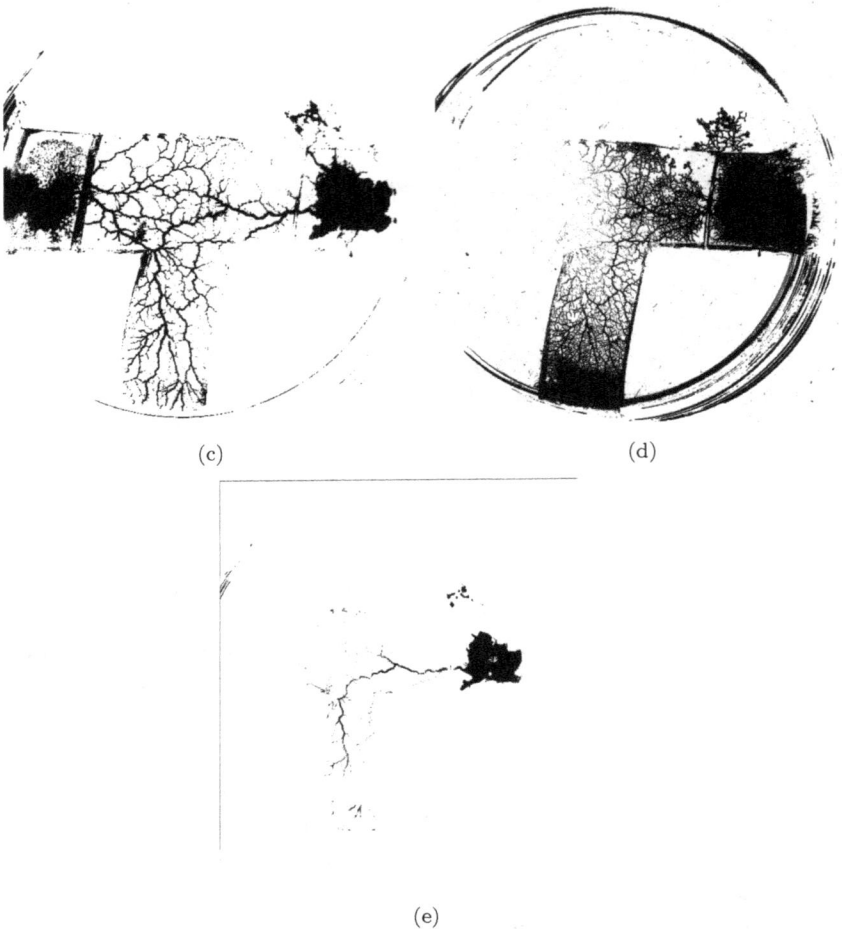

Fig. 14.10   *Continued.*

- to take in coloring from the closest colored oat flake;
- to mix two different colors to produce a third color;
- to transport color to a specified locus of a substrate.

The plasmodium demonstrates intelligent behavior by avoiding 'hazardous' domains of the environment when propagating towards target colors and by minimizing the length of the transport route between its source of origination and the target color.

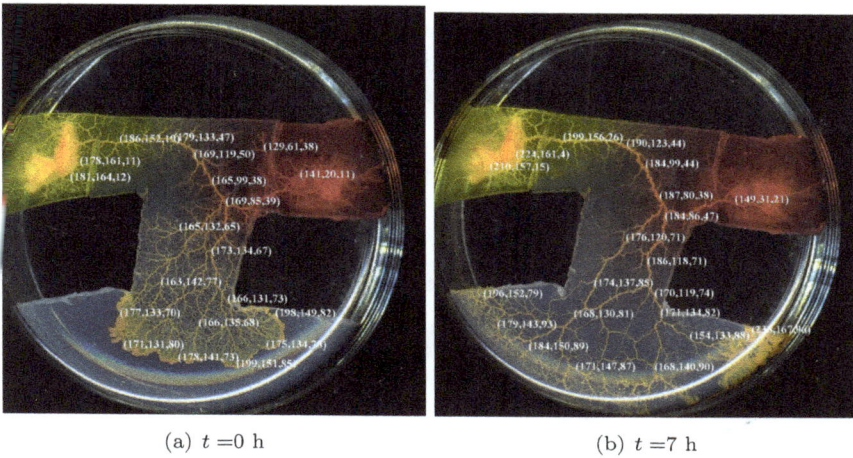

(a) $t = 0$ h   (b) $t = 7$ h

Fig. 14.11   Example of mixed color change during plasmodium's development.

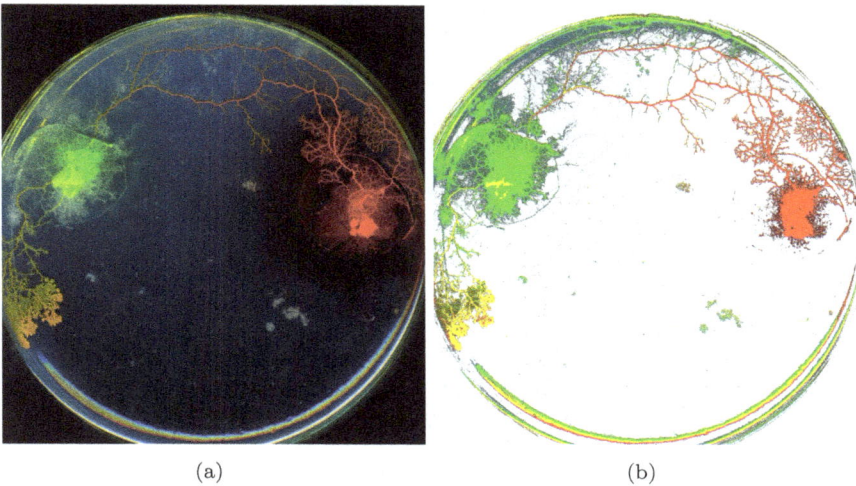

(a)   (b)

Fig. 14.12   Plasmodium mixes red and green colorings to produce yellow color: (a) scanned image of the experimental dish, (b) red, green and yellow colors are maximized; only red (c), green (d) and (e) yellow colored parts of the protoplasmic network are shown. Parameters of color separation are the same as used to produce Fig. 14.10b.

These findings manifest the plasmodium's potential for being a primary component for amorphous biorobotic devices. A plasmodium device transports substances by pumping the substance through the plasmodium's

(c)   (d)

(e)

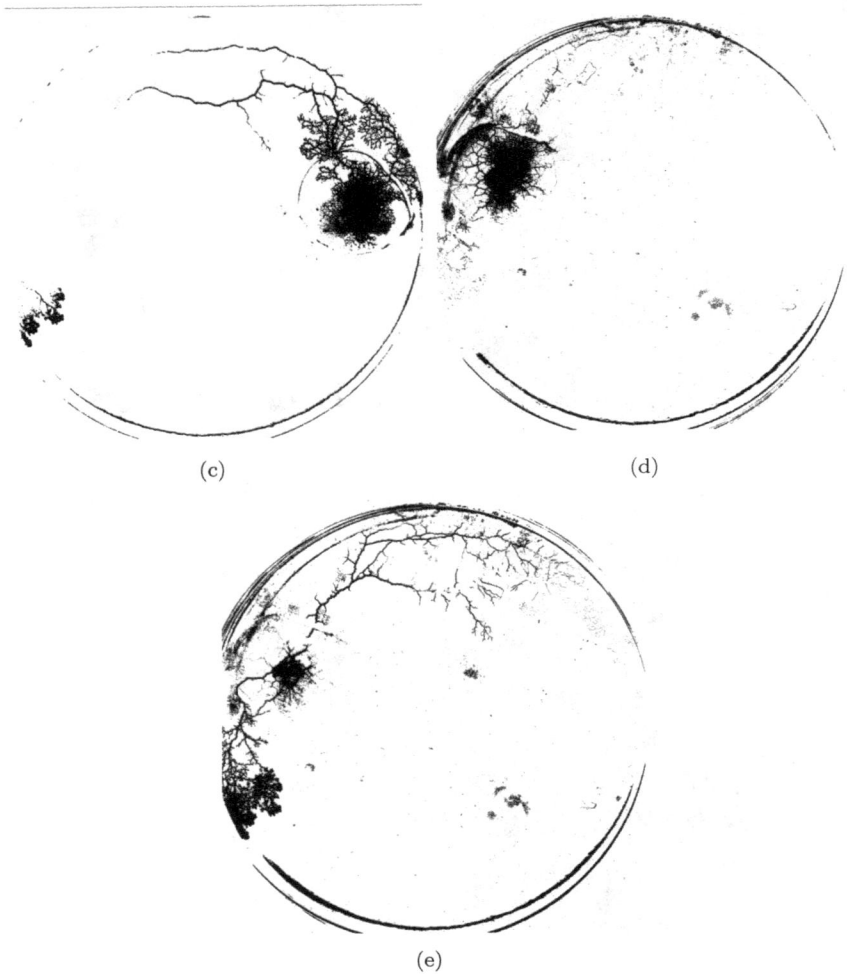

Fig. 14.12   *Continued.*

protoplasmic network and by direct relocation, or migration, of the whole plasmodium's body.

Colored substances are mixed in the plasmodium due to protoplasmic streaming. A streaming of protoplasm — and subsequent transfer of colored substances inside the protoplasm — is determined by propagation of contraction patterns in the plasmodium body. A spatio-temporal development of contraction patterns is dictated by traveling excitation waves,

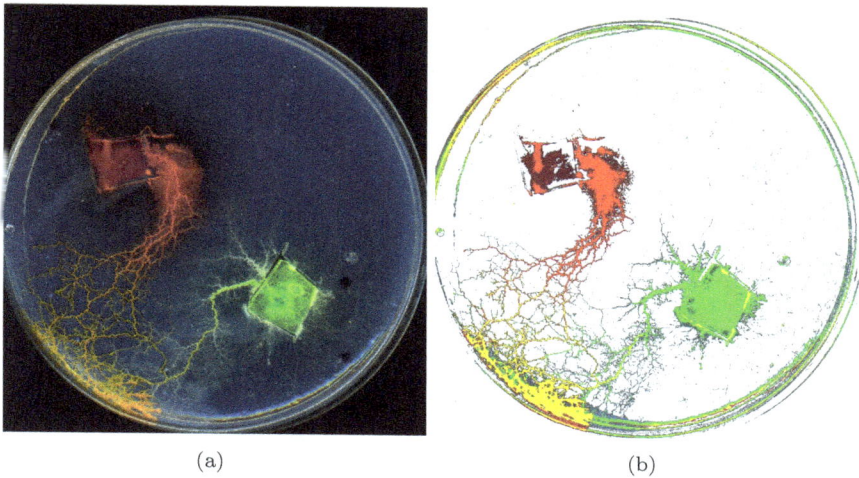

(a) (b)

Fig. 14.13 Controlling color mixing with repelling fields: (a) scanned image of plasmodia, positions of salt grains are marked by black disks on the bottom of Petri dish, (b) colors are separated. Separated color images are shown in (c) red, (d) green and (e) yellow.

generated by a source of biochemical oscillations.

We showed how to direct plasmodium towards pools of colored substances to be transported but never touched — a subject of controlling transportation of substances inside the plasmodium. The answer may lie in the experimental techniques of controlling protoplasmic streaming, e.g. applying thermal oscillations [Nakagaki et al. (2000)]. Nakagaki et al. [Nakagaki et al. (2000)] demonstrated that the rate and range of mixing of colored particles in plasmodium protoplasm depend on the current morphology of the plasmodium body. When plasmodium's morphology combines sheet-like and dendrite-tree-like structures, marked substances are well mixed and transported to all parts of the plasmodium's body. In situations when plasmodium exhibits a pronounced network of veins, the mixing is reduced and never occurs during one cycle of oscillations. Control of the morphology was achieved in [Nakagaki et al. (2000)] by modifying the chemical structure of the growth substrate. This may not be the best approach. The chemical structure of the substrate could not be dynamically changed during experiments. We believe that applying gradients of illumination would be the most practicable way of producing controllable spreading of substances inside the plasmodium's body.

(c)   (d)

(e)

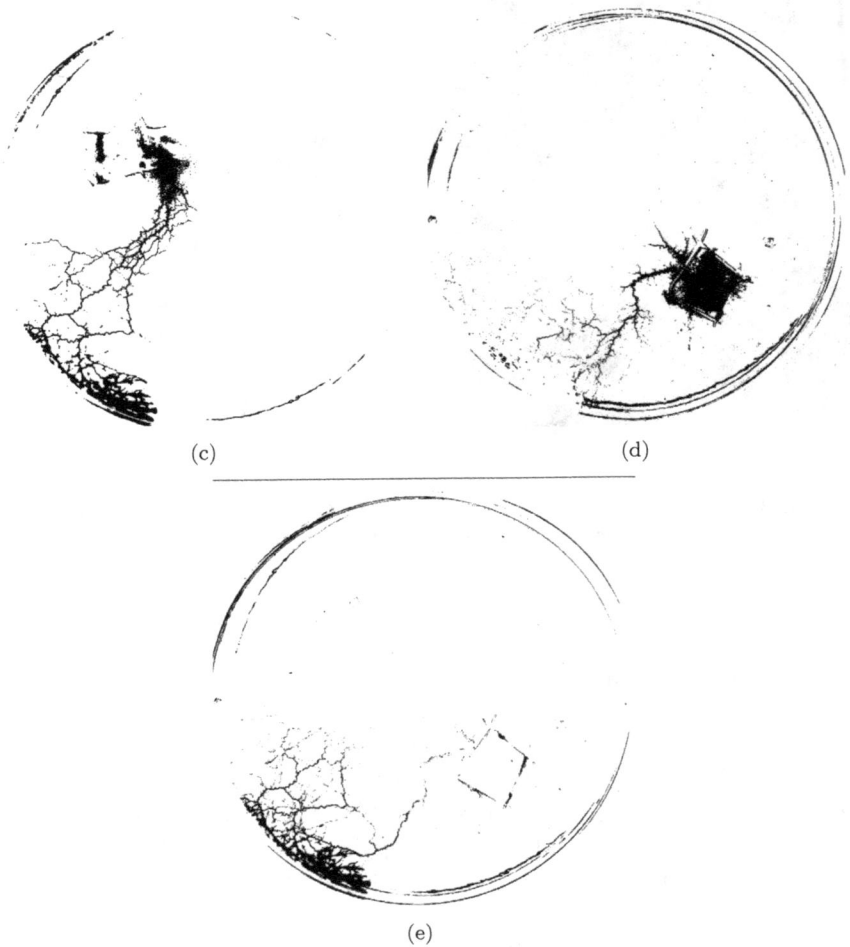

Fig. 14.13   *Continued.*

Our developments on transporting substances with plasmodium could also find their application in design of growing self-repairing electronic circuits. Given oat flakes impregnated with conductive substances, the plasmodium consumes the conductive material and distributes this material in its protoplasmic network. Electronic components, to be connected with 'plasmodium wires', are labeled with sources of nutrients.

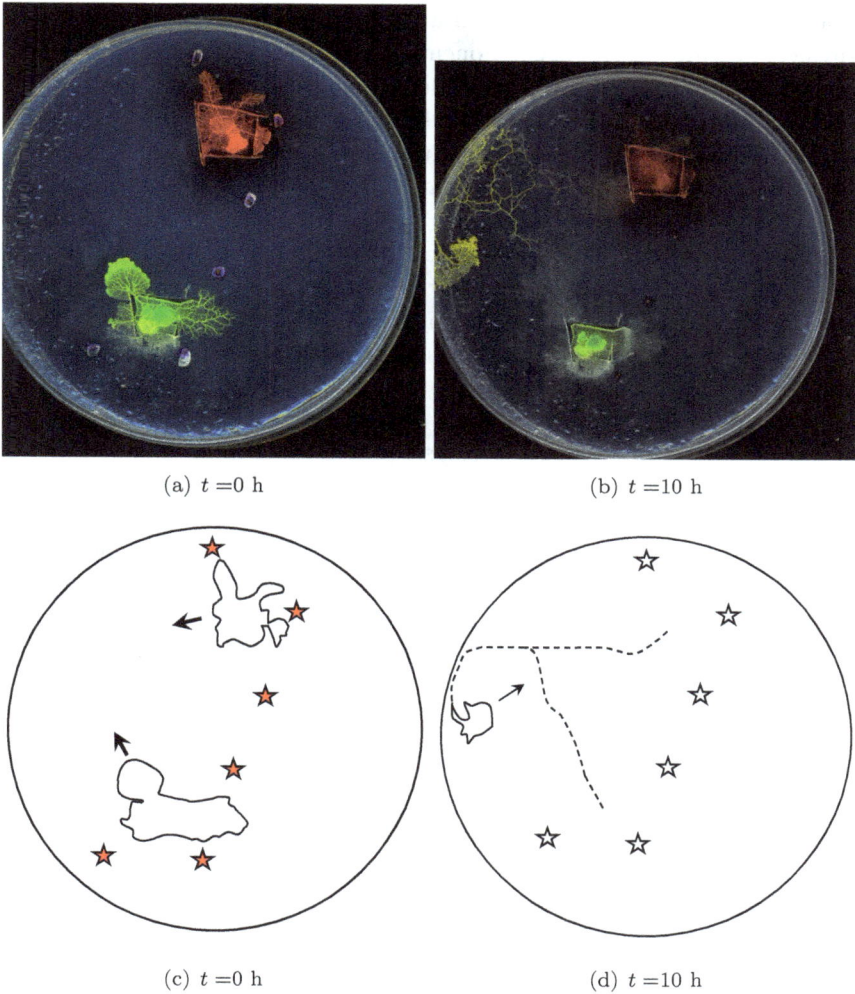

(a) $t = 0$ h   (b) $t = 10$ h

(c) $t = 0$ h   (d) $t = 10$ h

Fig. 14.14 Merging green- and red-colored plasmodia and forming autonomous yellow blob: (a) and (b) scanned images, (c) and (d) scheme of the manipulation, grains of salt are shown by stars, arrows represent vectors of plasmodia propagation.

Under guidance of attracting fields of nutrients, the plasmodium grows a minimum spanning tree or, if necessary, a proximity graph with cycles, connecting the electronic components. To fix the circuit built by plasmodium, we quickly decrease humidity. When drying, the plasmodium abandons the experimental area leaving behind the network of 'empty' protoplasmic tubes with the conductive material attached to walls of the tubes. Thus, wires connecting electronic components are formed.

Further studies are necessary to determine the exact nature of the distribution and rates of transportation of foreign substances inside the plasmodium network (which can be modulated by various environmental factors [Ueda (1980)]), and to investigate alternative methods of controlling the plasmodium's propagation.

The experimental findings presented in the chapter enlarged the range of controllable activities implemented by plasmodium. We found that in addition to programmable spanning of sources of attractants (Chap. 11.6) and manipulation of objects floating on a water surface, the plasmodium can propagate towards a target substance and transport the substance in a predetermined direction. These findings do not constitute a definite design of an amorphous robot but highlight basic principles of robot architecture — distributed sensing, decentralized decision making and parallel actuations — and show viable approaches to minimalist control of the robot behavior. Our findings suggest that possible engineering designs of amorphous robots will be based on principles of embedded control of morpho-functional machines [Yokoi et al. (2003)] implemented with electro-activated polymers impregnated with excitable reaction–diffusion chemical media [Melhuish et al. (2001); Murase et al. (2009); Yoshida et al. (2009)].

Chapter 15

Road planning with slime mould

Plasmodium of *P. polycephalum* approximates a shortest path, builds proximity graphs and grows spanning trees. Does the plasmodium emulate road networks? Given cities represented by oat flakes and plasmodium of *P. polycephalum* inoculated in one of the cities, will the plasmodium develop a protoplasmic network connecting oat flakes that matches the network of roads connecting the cities?

We started to look for an answer in 2006 [Adamatzky (2007)]. We arranged oat flakes in major cities of the United Kingdom from Birmingham to Portsmouth and from Penzance to Ramsgate. In all experiments plasmodium was inoculated in London (Fig. 15.1).

The plasmodium colonizes cities close to London (Fig. 15.1a) and then spreads itself further. In a day or two the plasmodium spans most cities in the south of England and disseminates as far as Exeter in the west (Fig. 15.1b).

Already at this initial stage of propagation the plasmodium networks mimic at least a few motorways, e.g. the segment of the M5 motorway from Gloucester to Taunton, and some segments of the M40, M1 and M25 motorways. An approximate location, shifted a bit south-westerly, of the Severn Bridge crossing the mouth of the Severn, is established by a protoplasmic tube connecting Bristol and Cardiff (Fig. 15.2).

In our early experiments [Adamatzky (2007)], we took too many cities into consideration, thus making it difficult to collect any reliable results. Also, we realized that it is unfeasible to keep boundaries of the experimental domain (south of England) open, because plasmodium tries to migrate outwards from the domain. Therefore, we decided to concentrate only on a few major urban areas and consider the whole United Kingdom as the plasmodium growth domain. Results of this adventure are described below.

(a)

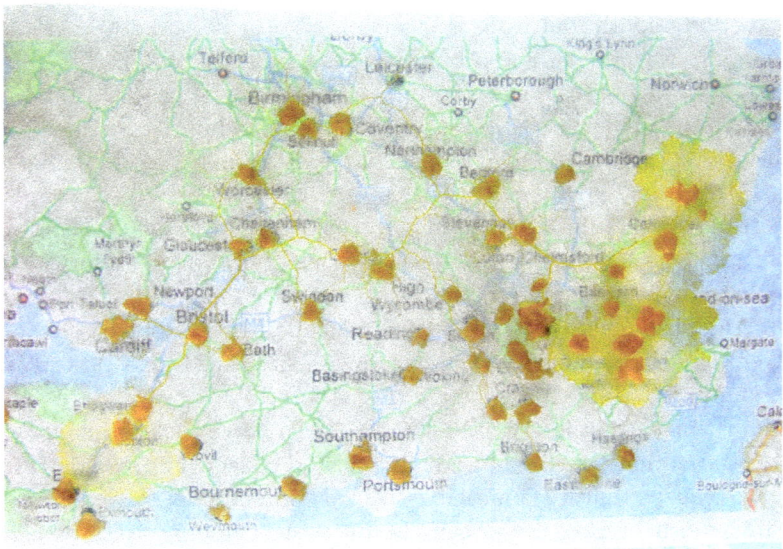

(b)

Fig. 15.1 Growth of plasmodium networks in south of England: (a) initial stage of network development, (b) plasmodium expands until it reaches Birmingham in the north, Brighton in the south and Exeter in the west. Plasmodium grows on non-nutrient agar gel.

Fig. 15.2 Magnified part of the south-west of England, from experiment shown in Fig. 15.1. Approximate location of Severn Bridge is imitated by a protoplasmic tube connecting oat flakes in Bristol and Cardiff. Segment of M5 motorway between Bristol and Exeter is marked by a thick protoplasmic tube.

15.1 United Kingdom in a gel

We use 120×120 mm^2 polystyrene square and 90 mm diameter round Petri dishes, and 2% agar gel (Select Agar, Sigma Aldrich) or a moistened filter paper as a non-nutrient growth substrate. Agar plates and filter papers are cut in a shape of the United Kingdom island.

We consider the 10 most populous urban areas in the United Kingdom (Fig. 15.3):

- Greater London,
- Bristol,
- Sheffield,
- Nottingham,
- Liverpool,
- Tyneside,
- Greater Glasgow,
- West Yorkshire,
- Greater Manchester,
- West Midlands,

Fig. 15.3   Schematic map of 10 most populous urban areas. Adapted from [Pointer (2005)].

as per the 2001 Census[1], see details and boundaries of the areas in [Pointer (2005)].

The areas are projected onto gel or filter paper and oat flakes are placed in the positions of the urban areas (Fig. 15.4). We try to match size of urban areas with size of oat flakes. At the beginning of each experiment, plasmodium is inoculated in the center of the Greater London urban area.

The Petri dishes with substrate and plasmodium are kept in the dark, at temperatures of 22–25°C, except for observation and image recording. Periodically, the dishes are scanned using an Epson Perfection 4490 scanner.

[1]Office for National Statistics, General Register Office for Scotland and Northern Ireland Statistics and Research Agency.

Fig. 15.4 Typical experimental setup: urban areas are represented by oat flakes, plasmodium is inoculated in London, the plasmodium spans oat flakes by protoplasmic transport network. Map of urban areas is adapted from [Pointer (2005)].

Large-scale disasters in urban areas are imitated by salt crystals[2] being placed in the center of a disaster zone.

Scanned images of dishes are enhanced for higher visibility: saturation is increased to 204 units, and contrast to 40 units. To ease readability of experimental images, we provide a complementary binary version of each image. The binarization is done as follows. Each pixel of a color image is assigned black color if the red R and green G components of its RGB color exceed some specified thresholds, $R > \theta_R$, $G > \theta_G$, and the blue component B does not exceed some threshold value $B < \theta_B$; otherwise, the pixel is assigned white color (exact values of the thresholds are indicated in the figure caption as $\Theta = (\theta_R, \theta_G, \theta_B)$.

---

[2]Saxa coarse grain sea salt, RHM Foods, Middlewich, CW10 0HD, UK.

15.2   Development of transport links

Being placed in the center of the Greater London urban area, plasmodium
typically consumes some nutrients from its nearest (London) oat flake and
starts propagating north, north-west or west (Fig. 15.5a and g). Birmingham (Fig. 15.5a and g) and Bristol (Fig. 15.6c and d) are usual candidates
which are spanned by London-originated plasmodium. When urban areas
in the Midlands are colonized by plasmodium and linked by protoplasmic
tubes, the plasmodium heads towards the Tyneside urban area (Fig. 15.5b–
h).

After taking in Tyneside, the plasmodium propagates north, crosses
Scottish boundaries and finally reaches the Glasgow urban area (Fig. 15.5c–
i). Then the plasmodium continues colonization of the substrate until all
urban areas (sources of nutrients) are colonized (Fig. 15.5d–f and j–l).

Examples of plasmodium networks connecting urban areas are shown
in Fig. 15.6. The figure demonstrates that, in general, the structure of the
network does not depend significantly on size and shape of the substrate
but mainly on the configuration of sources of nutrients. We demonstrated
this by undertaking experiments in a 90 mm round Petri dish (Fig. 15.6a
and b), 120 mm rectangular dishes fully covered with agar gel (Fig. 15.6c–f)
and the shape of the UK island cut out of an agar gel plate in a 120 mm
rectangular Petri dish (Fig. 15.6g and h).

The plasmodium does not stop its foraging activity even when all sources
of nutrients are colonized. It propagates away from the 'designated' area
(Fig. 15.6c–f) unless stopped by an unfriendly substrate (like the bottom
of a plastic Petri dish not covered by gel, Fig. 15.6g and h).

The plasmodium does not always keep all sources of nutrients spanned
by its protoplasmic tubes. Sometimes, a few tubes are abandoned during colonization. Thus, in Fig. 15.7, we see that at the beginning of its
development plasmodium links London and Nottingham (Fig. 15.7a and
c). When urban areas in the Midlands are colonized and linked by protoplasmic tubes to the Tyneside and Glasgow urban areas, the plasmodium
abandons its tube connecting the Greater London and Nottingham urban
areas (Fig. 15.7b and d).

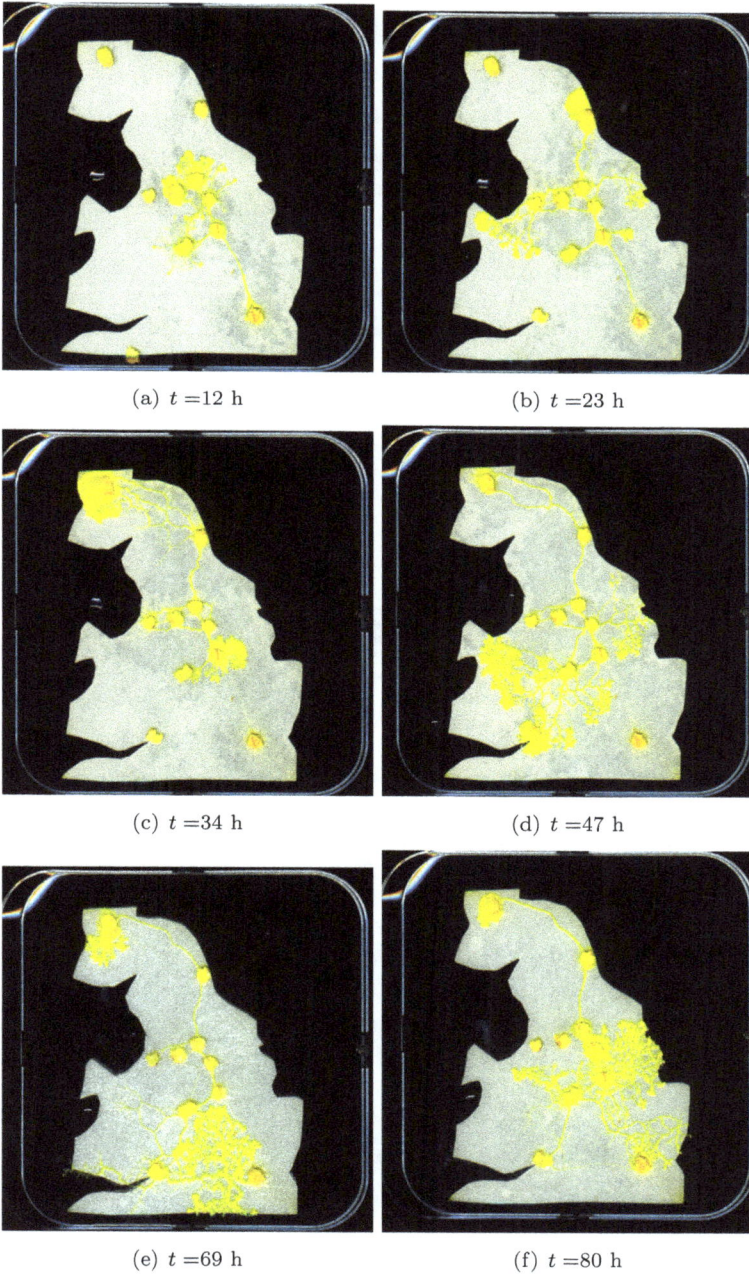

(a) $t = 12$ h   (b) $t = 23$ h

(c) $t = 34$ h   (d) $t = 47$ h

(e) $t = 69$ h   (f) $t = 80$ h

Fig. 15.5 Typical plasmodium development: (a)–(f) scanned images of experimental Petri dish; (g)–(l) binarized images, $\Theta = (200, 200, 150)$.

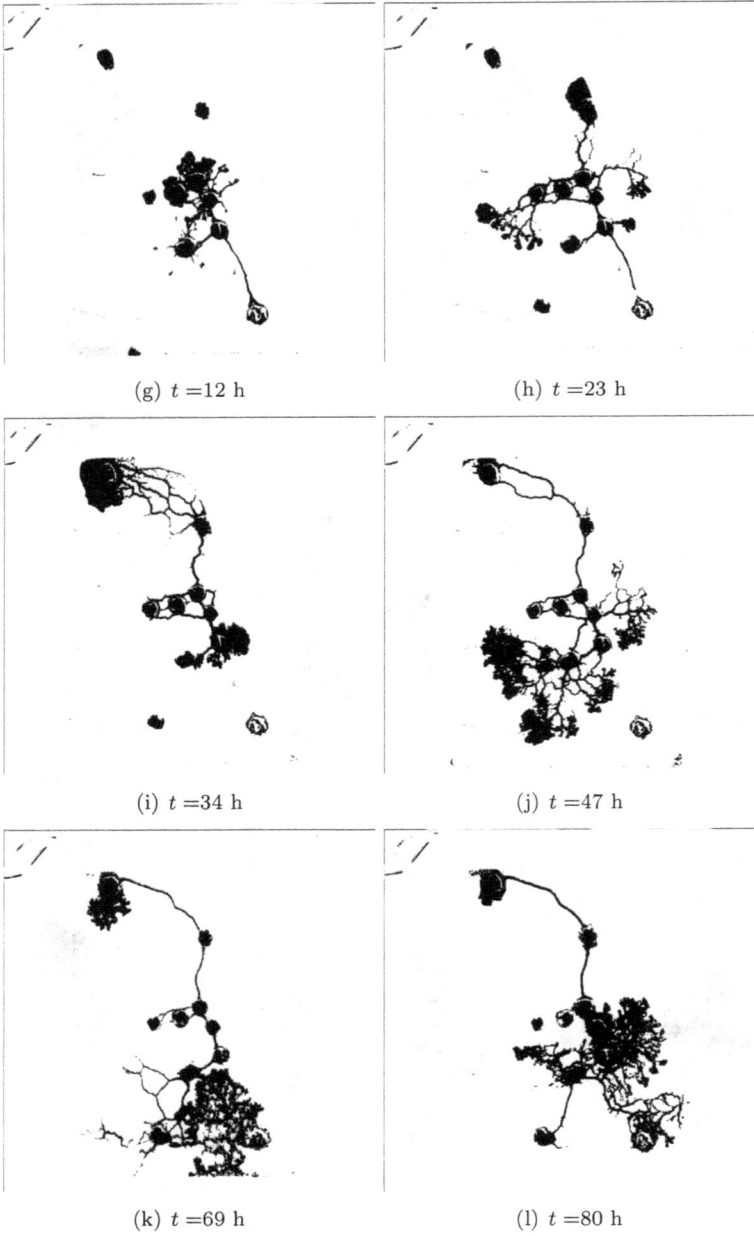

(g) $t = 12$ h

(h) $t = 23$ h

(i) $t = 34$ h

(j) $t = 47$ h

(k) $t = 69$ h

(l) $t = 80$ h

Fig. 15.5   *Continued.*

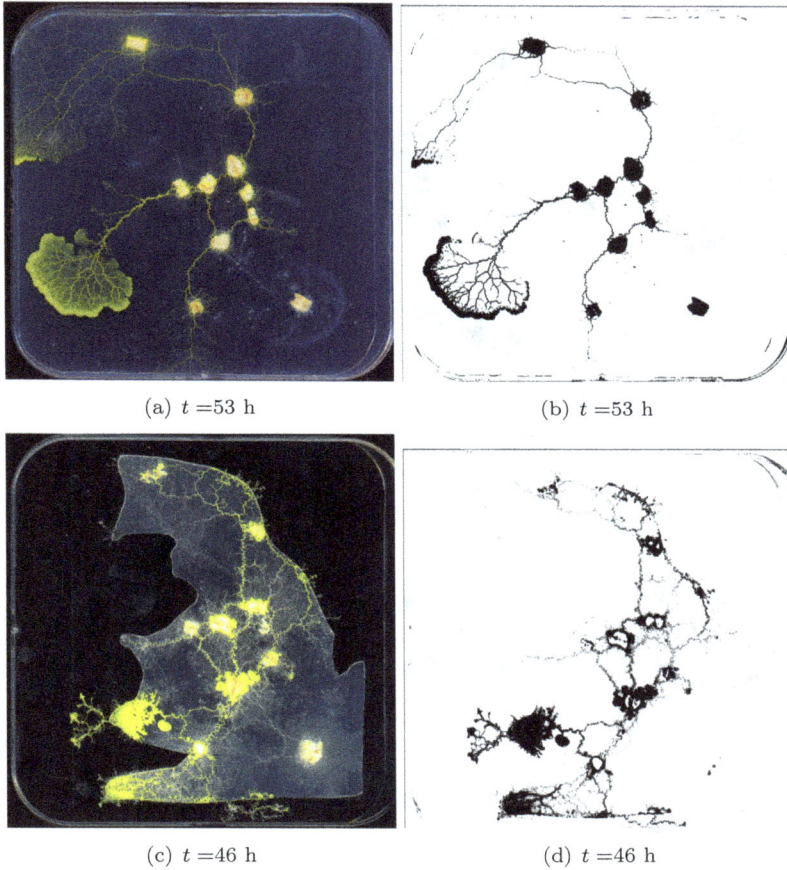

(a) $t = 53$ h   (b) $t = 53$ h

(c) $t = 46$ h   (d) $t = 46$ h

Fig. 15.6 Examples of protoplasmic networks occurring in experiments. Results of four independent experiments are shown in (a) and (b), (c) and (d), (e) and (f) and (g) and (h), respectively. Times passed after inoculation of plasmodium in London are shown in sub-figure caption. (a), (c), (e) and (g) are scanned images of experimental Petri dishes; (b), (d), (f) and (h) are binarized images. Thresholds of binarization are $\Theta = (170, 170, 100)$ for (b) and (d), and $\Theta = (130, 130, 100)$ for (f) and (h).

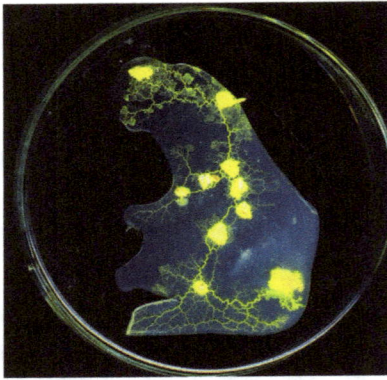

(e) $t = 63$ h

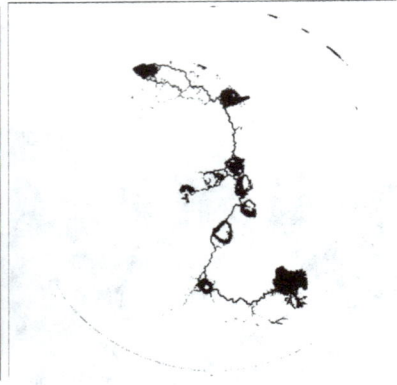

(f) $t = 63$ h

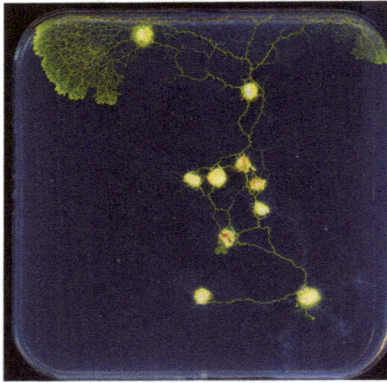

(g) $t = 47$ h

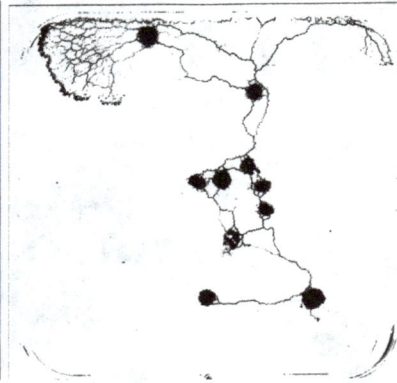

(h) $t = 47$ h

Fig. 15.6   *Continued.*

15.3   Weighted Physarum graphs

Even a limited number of examples (e.g. Figs. 15.6 and 15.7) demonstrate that plasmodium of *P. polycephalum* is a very dynamic system, whose morphology is continuously changing and whose spatio-temporal dynamics rarely reach a fixed stable point (unless humidity decreases and the plas-

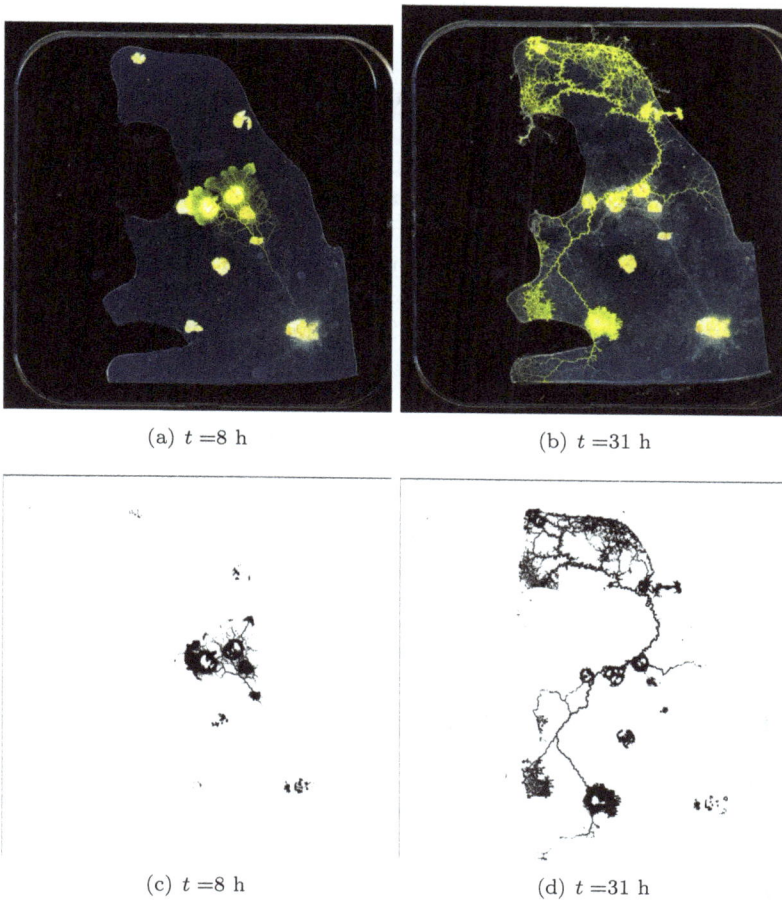

(a) $t = 8$ h

(b) $t = 31$ h

(c) $t = 8$ h

(d) $t = 31$ h

Fig. 15.7 Examples of reconfigurations of protoplasmic network: (a) and (b) scanned images, (c) and (d) binarized images, $\Theta = (140, 160, 100)$.

modium forms sclerotium). There is no such thing as a stationary configuration of a protoplasmic network; therefore, when extracting a generalized graph of transport links from plasmodium experiments we can only build a Physarum graph **P** as follows.

Let **U** be a set of the 10 most populous urban areas; $\mathbf{S} = \{S_1, \cdots, S_k\}$ is a set of the series S_i, $i = 1, \ldots, k$ (k is the number of experiments) of scanned images of plasmodium networks, $S_i = (s_i^1, \ldots, s_i^{m_i})$. For any two areas a and b from **U**, the weight of the edge (ab) is calculated as follows: $w(ab) = \sum_{S_i \in \mathbf{S}} \chi(S_i, a, b)$, where $\chi(S_i, a, b) = 1$ if there is at least one snapshot $s \in S_i$ which shows a protoplasmic tube connecting a and b. We do not take into account the exact configuration of the protoplasmic tubes but merely their existence. Each protoplasmic tube is counted just once for any particular experiment. We will also consider sub-graphs $\mathbf{P}_\alpha$ of **P**, $\alpha = 5, 10, 12$, defined as follows: for $a, b \in \mathbf{U}$, $(ab) \in \mathbf{P}_\alpha$ if $w(ab) > \alpha$.

Physarum graphs extracted from 25 laboratory experiments are shown in Fig. 15.8; the maximum edge weight is 22. The graph becomes planar when we remove edges with weights below 6 (Fig. 15.8b). The graph is acyclic, or a tree, when only edges appearing in over 40% of experiments are shown (Fig. 15.8d). If we increase the cut-off value to 14, the graph becomes disconnected, and the node corresponding to the Bristol urban area becomes isolated.

15.4   Physarum vs. Department for Transport

Let us check if there is any correspondence between transport links built by a Physarum machine and human-built motorways. We construct a motorway graph **M** as follows[3]. Let **U** be a set of the 10 most populous urban areas; for any two areas a and b from **U**, the nodes a and b are connected by an edge (ab) if there is a motorway starting in the vicinity of a and passing in the vicinity of b and not passing in the vicinity of any other urban area $c \in \mathbf{U}$.

The motorway graph **M** is shown in Fig. 15.9. By comparing **M** (Fig. 15.9) and the Physarum graph **P** and its sub-graphs (Fig. 15.8), we found the following.

Finding 24. *The motorway graph* **M** *is a sub-graph of the Physarum graph* **P**.

---

[3]The graph **M** is extracted from the motorway network as shown in `maps.google.com` and `www.openstreetmap.org`

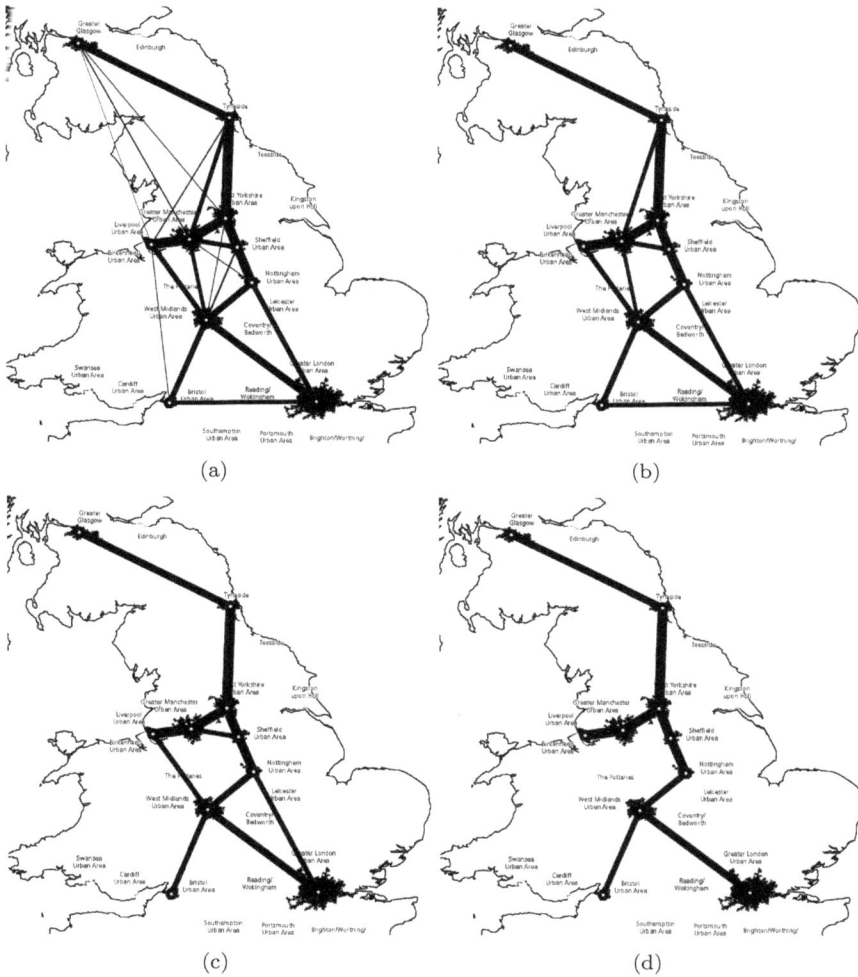

Fig. 15.8 Physarum graphs for various values of edge weights: (a) all edges of Physarum graph are shown, thickness of each edge is proportional to the edge's weight, (a)–(d) only edges with weights exceeding 5 (b), 10 (c) and 12 (d) are shown.

This shows that distributed 'logic' underpinning plasmodium's decision-making routines corresponds to human logic behind road-planning decisions. However, the graph **P** is not planar and thus poses little of practical importance. The motorway link M6/M74 connecting the Greater Manch-

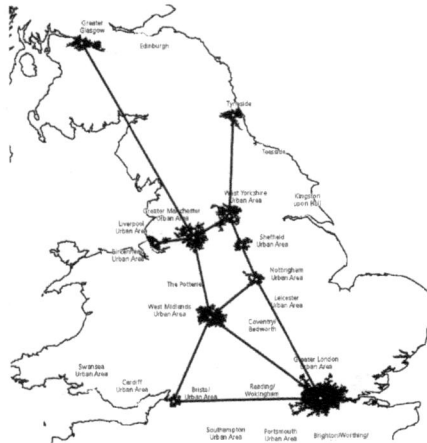

Fig. 15.9   Graph **M** of human-made motorway network connecting the 10 most populous urban areas.

ester and Greater Glasgow urban areas is represented by plasmodium only in three of 25 experiments. The corresponding edge does not appear in the graphs $\mathbf{P}_5$, $\mathbf{P}_{10}$ and $\mathbf{P}_{12}$ (Fig. 15.8b–d). The motorway M4, linking the Greater London and Bristol urban areas, is represented by protoplasmic tubes of *P. polycephalum* only in 20% of experiments, the graph $\mathbf{P}_5$ in Fig. 15.8b.

Finding 25.   *A Physarum machine satisfactorily approximates the motorway network linking the 10 most populous urban areas in the United Kingdom except for the motorway link M6/M74 connecting the Greater Manchester and Greater Glasgow urban areas.*

15.5   Proximity graphs and motorways

Is there a rationale behind plasmodium's behavior?

Let us have a look at the two most popular planar proximity graphs, the relative neighborhood graph [Toussaint (1980)] **RNG** (Fig. 15.10a) and the Gabriel graph [Gabriel and Sokal (1969); Matual and Sokal (1984)] **GG** (Fig. 15.10b) constructed over nodes corresponding to centers of urban areas. The graphs are related as **RNG** $\subseteq$ **GG** [Toussaint (1980); Matual and Sokal (1984); Jaromczyk and Toussaint (1992)]. Both graphs are important in spatial analysis and statistics, particularly **GG** which was

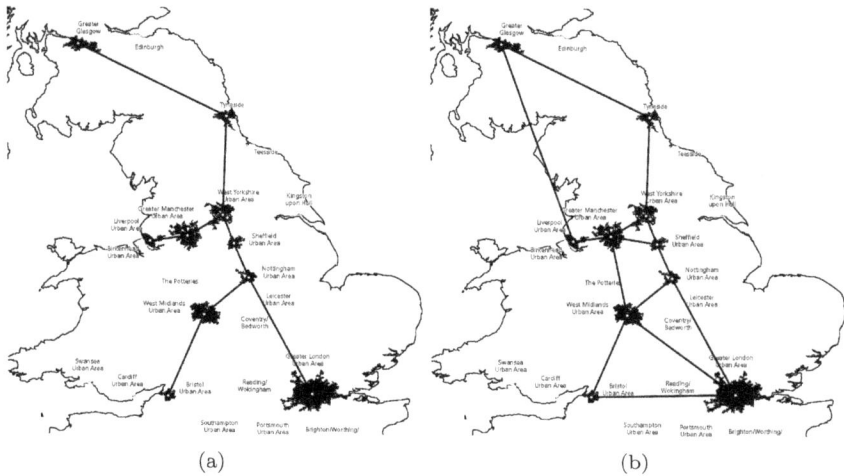

(a)   (b)

Fig. 15.10   Two proximity graphs constructed in urban areas **U**: (a) relative neighborhood graph **RNG**, (b) Gabriel graph **GG**.

invented specially to analyze and simulate the geographical distribution of biological populations [Gabriel and Sokal (1969)].

Our experiments have shown that

Finding 26. $\mathbf{P}_{10} \subset \mathbf{GG}$ *and* $\mathbf{P}_{12} = \mathbf{RNG}$.

Moreover, $\mathbf{P}_1 2$ is a minimum spanning tree (**MST**) over **U**. We know that $\mathbf{MST} \subseteq \mathbf{RNG}$ [Toussaint (1980)] but, in the particular case of urban areas' configurations **U**, we even have $\mathbf{MST(U)} = \mathbf{RNG(U)}$.

With regard to relations between the motorway graph and the proximity graphs, we see that **M** is neither a sub- nor a supergraph of **RNG** and **GG**. To transform **M** to **RNG**, one needs to remove two edges from and relocate one edge in **M**, while to transform **M** to **GG** no edges should be removed from but three edges added to **M**. The edge connecting Tyneside and Greater Glasgow is present in **RNG** and **GG** but absent in **M**.

Finding 27. *Experiments with plasmodium of* P. polycephalum *show that the M6/M74 motorway is not optimally positioned and should be rerouted from Newcastle to Glasgow. Alternatively, the M6/M74 motorway may remain intact but a new Newcastle–Glasgow motorway must be built.*

15.6   Imitating disasters

When making experiments, it is difficult to resist an impulse to imitate a large-scale disaster leading to contamination of one of the urban areas spreading to the surrounding areas. A disaster is imitated by placing a grain of salt in a substrate's locus corresponding to the urban area. Diffusion of sodium chloride in the substrate imitates progressive contamination of surrounding areas, making them temporarily uninhabitable.

Here we consider three examples: disaster in and contamination of the West Yorkshire urban area (Fig. 15.11, wet filter paper substrate, and Fig. 15.12, agar gel substrate) and the Tyneside urban area (Fig. 15.13, agar gel substrate).

In the experiment shown in Fig. 15.11, a plasmodium network spanning most urban areas is formed 32 h after inoculation of the Greater London area with plasmodium. We place a salt crystal in Leeds (West Yorkshire urban area) (Fig. 15.11a and d). In response to the increased concentration of sodium chloride the plasmodium abandons the West Yorkshire area and disconnects the area from neighboring Tyneside, Greater Manchester and Sheffield urban areas (Fig. 15.11b and e). At the same time the plasmodium starts mass exploration of Scotland and North Wales, and restores previously abandoned transport link with London (Fig. 15.11b and e). In around 27–30 h after the 'disaster' in West Yorkshire the plasmodium completes its evacuation from the Midlands and northern England and regroups itself in northern Scotland and southern England (Fig. 15.11c and f).

Due to the lower rate of sodium chloride diffusion in agar gel (compared to wet filter paper), the plasmodium's response to imitated contamination of agar gel is less dramatic.

As in the previous experiment, we wait until plasmodium forms a well-established protoplasmic network and then place a salt crystal in Leeds (Fig. 15.12a and d). The plasmodium temporarily breaks all transport links leading to the contaminated zone (West Yorkshire), and increases exploratory activity in Wales (even developing pronounced protoplasmic routes in west Wales) and the south-west of England (Fig. 15.12b and e). One and half days after contamination of West Yorkshire the plasmodium restores transport links with Leeds (Fig. 15.12c and f) and decreases its activity in Wales and the south-west.

In the example shown in Fig. 15.13, we strike an urban area with contaminant before any transport link leading to the area is established. Plasmodium inoculated in the Greater London area spans urban areas in the

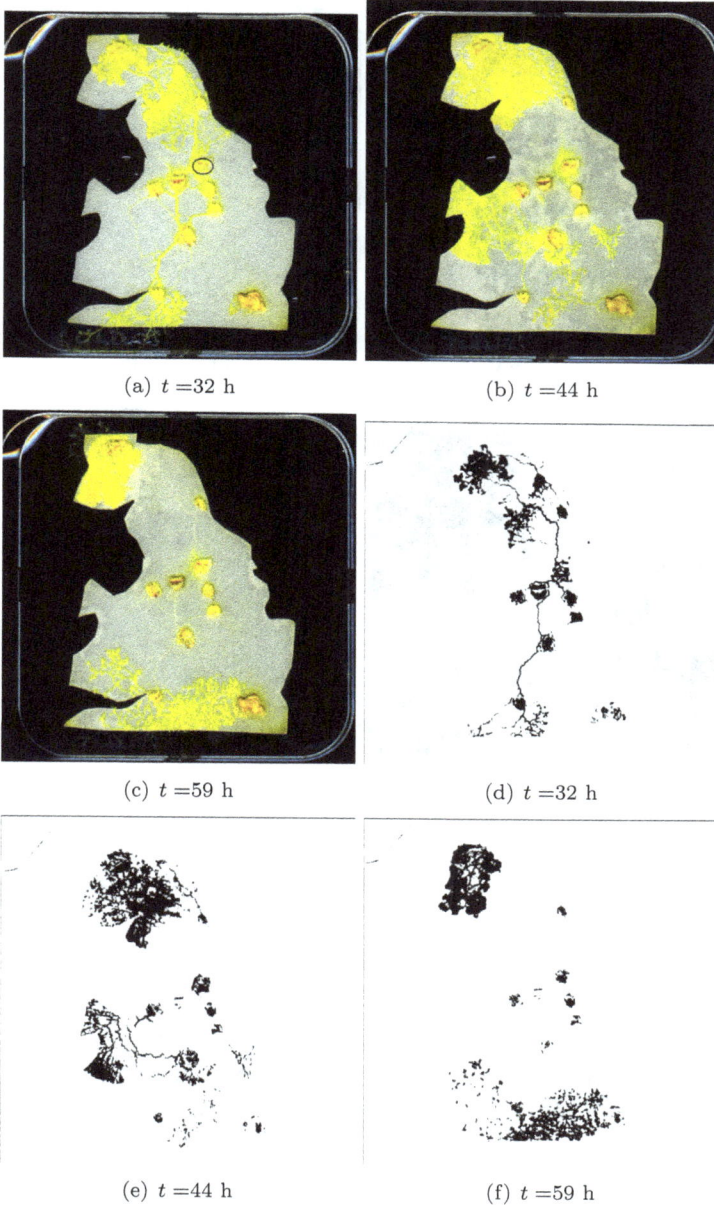

(a) $t = 32$ h

(b) $t = 44$ h

(c) $t = 59$ h

(d) $t = 32$ h

(e) $t = 44$ h

(f) $t = 59$ h

Fig. 15.11   Plasmodium's response to disaster in and subsequent contamination of West Yorkshire area, filter paper substrate: (a)–(c) scanned images, (d)–(f) binarized images, $\Theta = (200, 200, 20)$. Grain of salt placed (marked by circle in (a)) in West Yorkshire urban area at 32 h of plasmodium development. Images (b)–(c) and (e)–(f) show the plasmodium's response to increased concentration of sodium chloride.

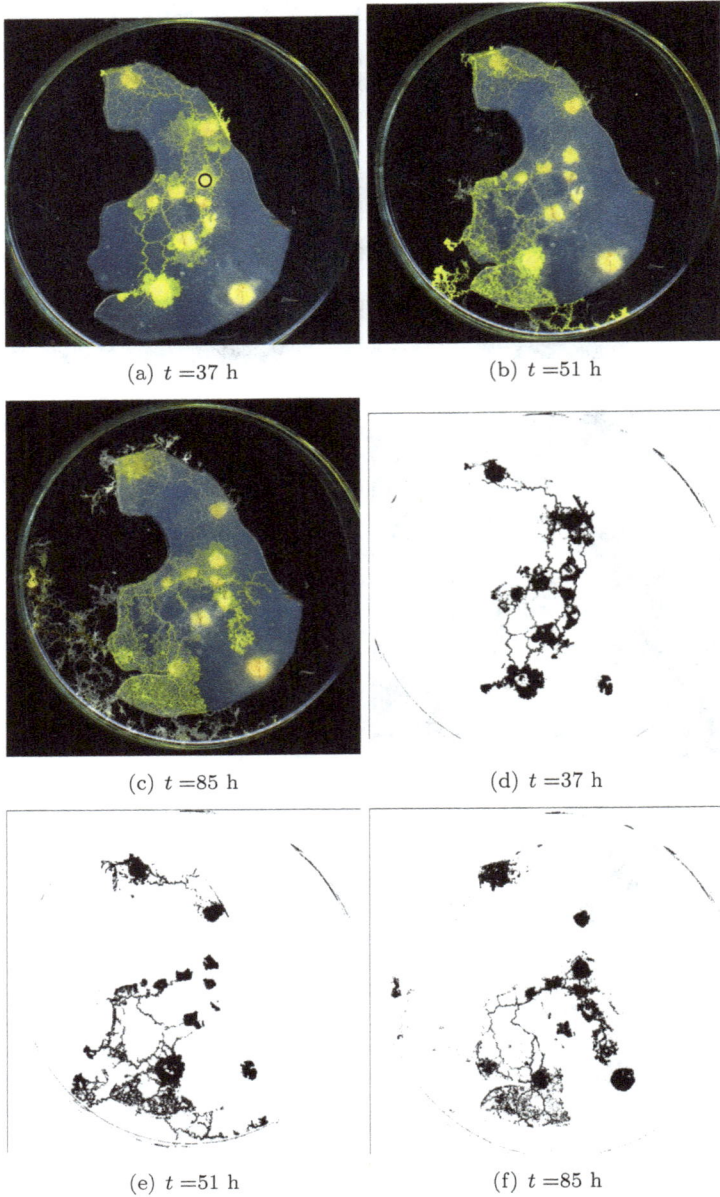

(a) $t =37$ h   (b) $t =51$ h

(c) $t =85$ h   (d) $t =37$ h

(e) $t =51$ h   (f) $t =85$ h

Fig. 15.12   Plasmodium's response to disaster in and subsequent contamination of West Yorkshire area, agar gel substrate: (a)–(c) scanned images, (d)–(f) binarized images, $\Theta = (200, 200, 20)$; a grain of salt placed (marked by circle in (a)) in West Yorkshire urban area at 37 h of plasmodium development. Images (b)–(c) and (e)–(f) show the plasmodium's reaction to increased concentration of sodium chloride.

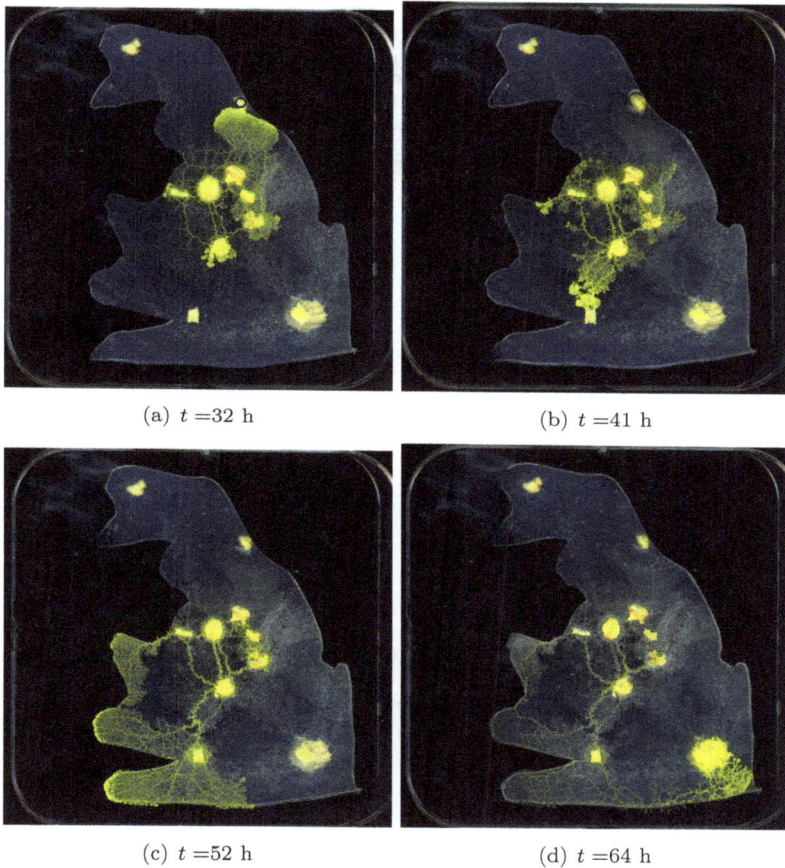

(a) $t = 32$ h

(b) $t = 41$ h

(c) $t = 52$ h

(d) $t = 64$ h

Fig. 15.13 Plasmodium's response to disaster in and subsequent contamination of Tyneside area, agar gel substrate, (a)–(d) scanned images, (e)–(h) binarized images, $\Theta = (200, 200, 20)$; a grain of salt placed, marked by circle in (a), in Tyneside urban area at 32 h of plasmodium development. Images (b) and (c) and (e) and (f) show the plasmodium's reaction to increased concentration of sodium chloride.

Midlands with a protoplasmic network (Fig. 15.13a and e). When the plasmodium's active zone approaches Tyneside, we place a salt crystal in Newcastle (Fig. 15.13a and e). In response to the contamination, the plasmodium abandons its attempt to colonize northern England and Scotland but heads its foraging activity west and south, and 9 h after contamination of Tyneside the plasmodium reaches Bristol (Fig. 15.13b and f). The plasmodium increases its exploration of Wales and the south-west of England

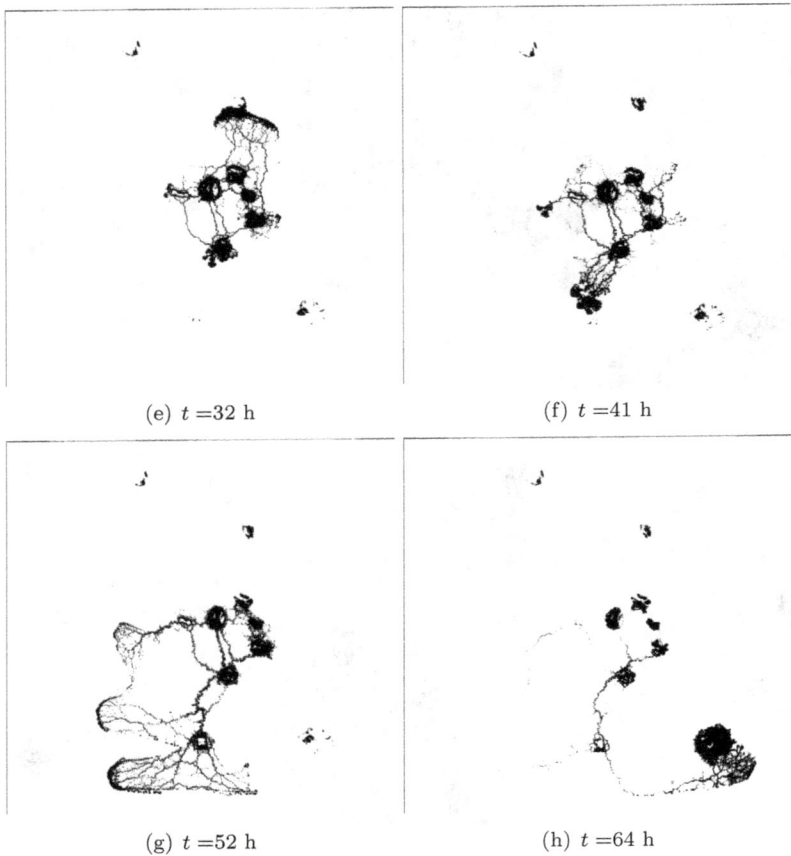

(e) $t = 32$ h

(f) $t = 41$ h

(g) $t = 52$ h

(h) $t = 64$ h

Fig. 15.13 *Continued.*

(Fig. 15.13c and g). In 32 h after the disaster strikes Tyneside the plasmodium restores the transport network between the Midlands and London, this time via Bristol (Fig. 15.13c and g).

Finding 28. *Contamination of a single urban area stimulates exploration of uncolonized areas and leads to restoration of previously abandoned transport links.*

15.7 Summary

We represented the 10 most populated UK urban areas by sources of nutrients, inoculated plasmodium of *P. polycephalum* in one of the areas and analyzed dynamics of colonization of the areas by the plasmodium. We studied space–time dynamics of spanning the urban areas by the plasmodium's network of protoplasmic tubes and demonstrated that the plasmodium transport network sufficiently well matches the topology of the existing human-made motorway networks. We found two discrepancies.

The M4 motorway (Bristol–London) rarely occurs in plasmodium networks. The route M6/M74 is absent.

As a 'byproduct' of the experiments, we provided an insight into bioinspired response to disastrous contamination of urban areas. Two main components of the response are (1) exploring uncolonized territories and (2) restoration of abandoned transport links.

Our experiments did not take terrain into account. This may explain the 'anomalous' situation with plasmodium not imitating the route M6/M74 but developing a transport link directly from Newcastle to Glasgow. Terrain-based experiments on Physarum road planning may form a subject for further studies.

The Physarum-based approach to evaluation of road networks can be easily expanded to any country (Fig. 15.14). The only thing you need is to take a page from your road atlas or print a map from `maps.google.com`. Spray the paper with water, wet but not soaking wet. Position oat flakes on top of cities and put a piece of plasmodium in any place on the map. In one or two days you will see a sophisticated network of protoplasmic tubes developed on the map (Fig. 15.15a). By making a few experiments, you can collect statistics and conclude whether the plasmodium well imitates the road network of your country or not (Fig. 15.14). If left to dry in a dark place the plasmodium does not fructify but forms a sclerotium, which, if you are lucky enough, will represent a 'frozen' replica of the existing roads (Fig. 15.15b).

(a)

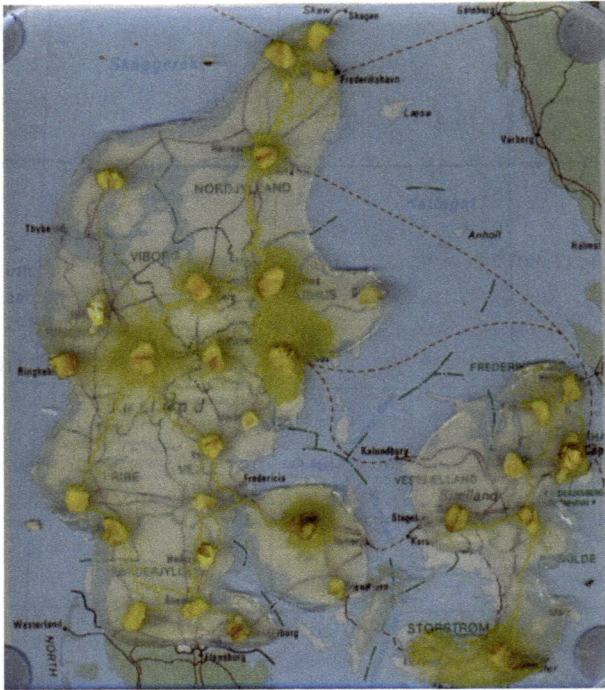

(b)

Fig. 15.14   Images of scoping experiments on Physarum approximating road networks in USA (a) and Denmark (b). Oat flakes are placed in locations of major cities. (a) Plasmodium is inoculated in New York, (b) plasmodium is inoculated in Copenhagen, Odense and Aarhus. Images are scanned 3 days after initial inoculation.

(a)

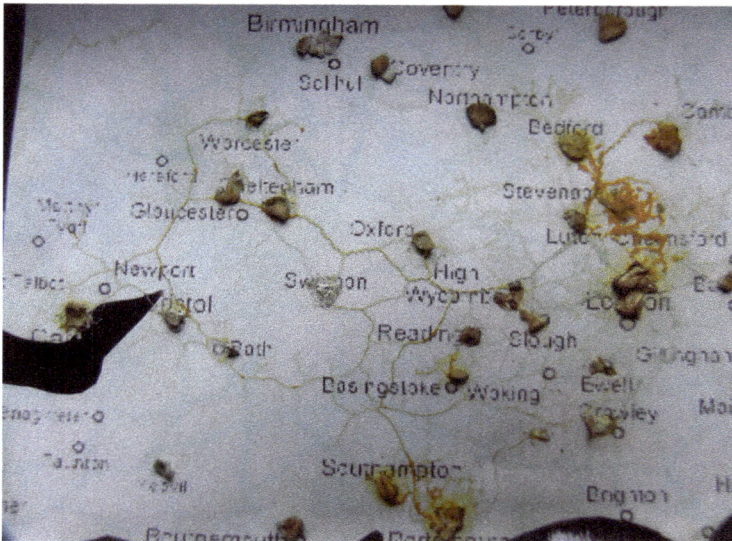

(b)

Fig. 15.15 Replicating road networks with plasmodium growing on a wet paper: (a) photograph of live protoplasmic network, (b) dried plasmodium, sclerotium, represents results of the experiment.

Epilogue

A Physarum machine is a programmable amorphous biological computer experimentally implemented in plasmodium of *Physarum polycephalum*.

A Physarum machine on a nutrient-rich substrate behaves as an auto-wave in an excitable medium. On a non-nutrient substrate it propagates similarly to a wave fragment in a sub-excitable medium.

A Physarum machine can be programmed by configurations of repelling (salt) and attracting (food) gradients, and localized reflectors (illuminated obstacles).

A Physarum machine solves mazes. It represents a path from start to finish sites in a maze by its protoplasmic tube.

A Physarum machine approximates a planar Voronoi diagram. Data planar points are mapped by pieces of plasmodia. Bisectors of the Voronoi diagram are composed of the substrate's loci not colonized by plasmodium.

A Physarum machine computes a nearest-neighborhood graph, a spanning tree, a relative neighborhood graph and a Delaunay triangulation at various stages of its development. Nodes of a graph are represented by sources of nutrients, edges by protoplasmic tubes connecting the sources.

A Physarum machine is a universal computer. The machine implements Boolean logic conjunction, disjunction and negation on a geometrically constrained substrate. The machine can also realize binary adders.

A Physarum machine is a biological implementation of a Kolmogorov–Uspensky machine. It can therefore be considered as a biological prototype of all storage modification machines and modern computer architectures.

A Physarum machine is a programmable manipulator. The machine can push and pull objects floating on a water surface by expanding and contracting its protoplasmic tubes.

A Physarum machine can act as a miniature boat motor. If attached to a tiny object floating on a water surface, the plasmodium oscillates in the water and propels the floater forward.

A Physarum machine can act as an adaptive transporter of liquids and micro-particles. The transportation is programmed by configurations of repelling and attracting fields.

A Physarum machine can evaluate optimality of human-made road networks. If cities are represented by sources of nutrients, the machine represents roads, connecting the cities, as protoplasmic tubes.

A Physarum machine is very cheap to make and easy to maintain. The machine functions on a wide range of substrates and in a broad scope of environmental conditions.

A Physarum machine is a 'green' and environmentally friendly unconventional computer. It requires little energy to sustain and can be fed almost any kind of nutrients. The unwanted machine can be stored as sclerotium or returned to a food chain.

Bibliography

Achenbach U. and Wolfarth-Bottermann K. E. Oscillating contractions in protoplasmic strands of *Physarum*. J. Exp. Biol. 85 (1980) 21–31.

Achenbach F. and Weisenseel M. H. Ionic currents traverse the slime mould
Physarum. Cell Biol. Int. Rep. 5 (1981) 375–379.

Adamatzky A. Neural algorithm for constructing minimal spanning tree. Neural
Network World 6 (1991) 335–339.

Adamatzky A. Reaction–diffusion algorithm for constructing discrete Voronoi diagram. Neural Network World 6 (1994) 635–643.

Adamatzky A. Reaction–diffusion computer: massively parallel molecular computation. Math. Res. 96 (1996) 287–290.

Adamatzky A. and Tolmachiev D. Chemical processor for computation of skeleton
of planar shape. Adv. Mater. Opt. Electron. 7 (1997) 135–139.

Adamatzky A. and Holland O. Reaction–diffusion and ant-based load balancing
of communication networks. Kybernetes 31(5) (2002) 667–681.

Adamatzky A. Computing in Non-linear Media and Automata Collectives (IoP
Publishing, Bristol, 2001).

Adamatzky A. and Melhuish C. Phototaxis of mobile excitable lattices. Chaos
Solitons Fractals 13 (2002) 171–184.

Adamatzky A. and De Lacy Costello B. P. J. Experimental logical gates in a
reaction–diffusion medium: the XOR gate and beyond. Phys. Rev. E 66
(2002) 046112.

Adamatzky A. and De Lacy Costello B. P. J. Collision-free path planning in the
Belousov–Zhabotinsky medium assisted by a cellular automaton. Naturwissenschaften 89 (2002) 474–478.

Adamatzky A. (ed.) Collision-Based Computing (Springer, London, 2003).

Adamatzky A. Collision-based computing in Belousov–Zhabotinsky medium.
Chaos Solitons Fractals 21 (2004) 1259–1264.

Adamatzky A., De Lacy Costello B., Melhuish C., and Ratcliff N. Experimental
implementation of mobile robot taxis with onboard Belousov-Zhabotinsky
chemical medium. Mater. Sci. Eng. C 24 (2004) 541–548.

Adamatzky A., De Lacy Costello B., and Asai T. Reaction–Diffusion Computers
(Elsevier, Amsterdam, 2005).

Adamatzky A., Skachek S., De Lacy Costello B., and Melhuish C. Manipulating objects with chemical waves: open loop case of experimental Belousov–Zhabotinsky medium. Phys. Lett. A 350 (2005) 201–209.

Adamatzky A. and Teuscher C. From Utopian to Genuine Unconventional Computers (Luniver Press, Beckington, 2006).

Adamatzky A. Physarum machines: encapsulating reaction–diffusion to compute spanning tree. Naturwissenschaften 94 (2007) 975–980.

Adamatzky A. Growing spanning trees in plasmodium machines. Kybernetes: Int. J. Syst. Cybern. 37 (2008) 258–264.

Adamatzky A. Physarum machines: encapsulating reaction–diffusion to compute spanning tree. Naturwissenschaften 94 (2007) 975–980.

Adamatzky A. Physarum machine: implementation of a Kolmogorov–Uspensky machine on a biological substrate. Parallel Process. Lett. 17 (2007) 455–467.

Adamatzky A. From reaction–diffusion to Physarum computing. Invited talk at Los Alamos Lab workshop 'Unconventional Computing: Quo Vadis?', Santa Fe, NM, March 2007.

Adamatzky A. and De Lacy Costello B. Binary collisions between wave-fragments in a sub-excitable Belousov–Zhabotinsky medium. Chaos Solitons Fractals 34 (2007) 307–315.

Adamatzky A., Bull L., De Lacy Costello B., Stepney S., and Teuscher C. Unconventional Computing 2007 (Luniver Press, Beckington, 2007).

Adamatzky A. Developing proximity graphs by Physarum Polycephalum: does the plasmodium follow Toussaint hierarchy? Parallel Process. Lett. 19 (2008) 105–127.

Adamatzky A., De Lacy Costello B., and Shirakawa T. Universal computation with limited resources: Belousov–Zhabotinsky and Physarum computers. Int. J. Bifurcat. Chaos 18 (2008) 2373–2389.

Adamatzky A. and Jones J. Towards Physarum robots: computing and manipulating on water surface. J. Bionic Eng. 5 (2008) 348–357.

Adamatzky A. If BZ medium did spanning trees these would be the same trees as *Physarum* built. Phys. Lett. A 373 (2009) 952–956.

Adamatzky A. Hot ice computer. Phys. Lett. A 374 (2009) 264–271.

Adamatzky A. and Jones J. Road planning with slime mould: if Physarum built motorways it would route M6/M74 through Newcastle. Int. J. Bifurcat. Chaos (2009).

Adamatzky A. and Jones J. Programmable reconfiguration of Physarum machines. Nat. Comput. (2009).

Adamatzky A. *Physarum* boats: if plasmodium sailed it would never leave a port. Appl. Bionics Biomech. 7 (2010) 31–39.

Adamatzky A. Routing *Physarum* with repellents. Eur. Phys. J. E 31 (2010) 403–410.

Agladze K., Magome N., Aliev R., Yamaguchi T., and Yoshikawa K. Finding the optimal path with the aid of chemical wave. Physica D 106 (1997) 247–254.

Alber J., Dorn F., and Niedermeier R. Experiments on Optimally Solving NP-complete Problems on Planar Graphs, manuscript (2001).http://www.ii.uib.no/~frederic/ADN01.ps

Allen R. D., Pitts W. R., Speir D., and Brault J. Shuttle-streaming: synchronization with heart production in slime mold. Science 142 (1963) 1485–1487.

Anderson O. R. A fine structure study of Physarum polycephalum during transformation from sclerotium to plasmodium: a six-stage description. J. Eukaryotic Microbiol. 39 (2007) 213–215.

Anton F. In Proc. Sixth IEEE Int. Symp. Voronoi Diagrams, Copenhagen, Denmark, 2009.

Aono M. and Gunji Y.-P. Resolution of infinite-loop in hyperincursive and non-local cellular automata: introduction to slime mold computing. Computing Anticipatory Systems, AIP Conf. Proc. 718 (2001) 177–187.

Aono M. and Gunji Y.-P. Material implementation of hyperincursive field on slime mold computer. Computing Anticipatory Systems, AIP Conf. Proc. 718 (2004) 188–203.

Aono M. and Hara M. Amoeba-based nonequilibrium neurocomputer utilizing fluctuations and instability. Lect. Notes Comput. Sci. 4618 (2007) 41.

Bardzin' J. M. On universality problems in theory of growing automata. Dokl. Akad. Nauk SSSR 157 (1964) 542–545.

Barzdin' J. M. and Kalnins J. A universal automaton with variable structure. Autom. Control Comput. Sci. 8 (1974) 6–12.

Beato V. and Engel H. Pulse propagation in a model for the photosensitive Belousov–Zhabotinsky reaction with external noise. In Noise in Complex Systems and Stochastic Dynamics, ed. by Schimansky-Geier L., Abbott D., Neiman A., and Van den Broeck C. (Proc. SPIE 5114, Bellingham) (2003), pp. 353–362.

Beckers R., Goss S., Deneubourg J. L., and Pasteels J. M. Colony size, communication and ant foraging strategy. Psyche 96 (1989) 239–256.

Bialczyk J. An action spectrum for light avoidance by Physarum Nudum plasmodia. Photochem. Photobiol. 30 (1979) 301–303.

Blackwell M., Waa J. V., and Reynolds M. Survival of myxomycetes sclerotia after exposure to high temperature. Mycologia 76 (1984) 752–754.

Blass A. and Gurevich Y. Algorithms: a quest for absolute definitions. Bull. Eur. Assoc. TCS 81 (2003) 195–225.

Block I. and Wohlfarth-Bottermann K. E. Blue light as a medium to influence oscillatory contraction frequency in physarum. Cell Biol. Int. Rep. 5 (1981) 73–81.

Burgin M. Inductive Turing machines. Not. Acad. Sci. USSR 270 (1983) 1289–1293.

Burgin M. Arithmetic hierarchy and inductive Turing machines. Not. Acad. Sci. USSR 299(3) (1988) 530–533.

Bykov A. V., Priezzhev A. V., Lauri J., and Myllyl'a R. Doppler OCT imaging of cytoplasm shuttle flow in Physarum polycephalum. J. Biophoton. 2(8) (2009) 540–547.

CalPhotos, a project of BSCIT, University of California, Berkeley. http://calphotos.berkeley.edu/

Calude C. S., Dinneen M. J., Paun G., Rozenberg G., and Stepney S. In 5th Int. Conf. Unconventional Computation (Springer, New York, 2006).

Cartigny J., Ingelrest F., Simplot-Ryl D., and Stojmenovic I. Localized LMST and RNG based minimum-energy broadcast protocols in ad hoc networks. Ad Hoc Networks 3 (2005) 1-16.

Cheng G.-X., Ikegami M., and Tanaka M. A resistive mesh analysis method for parallel path searching. In Proc. 34th Midwest Symp. Circuits and Systems, 1991, vol. 2, pp. 827–830.

Chet I. and Henis Y. Sclerotial morphogenesis in fungi. Annu. Rev. Phytopathol. 13 (1975) 169–192.

Chong F. Analog techniques for adaptive routing on interconnection networks. MIT Transit Note No. 14, 1993.

Cloteaux B. and Rajan D. Some separation results between classes of pointer algorithms. In Proc. Eighth Workshop Descriptional Complexity of Formal Systems (DCFS '06), 2006, pp. 232–240.

De Lacy Costello B. P. J. Constructive chemical processors — experimental evidence that shows that this class of programmable pattern forming reactions exist at the edge of a highly non-linear region. Int. J. Bifurcat. Chaos 13 (2003) 1561–1564.

De Lacy Costello B. and Adamatzky A. On multitasking in parallel chemical processors: experimental findings. Int. J. Bifurcat. Chaos 13 (2003) 521–533.

De Lacy Costello B. P. J., Hantz P., and Ratcliffe N. M. Voronoi diagrams generated by regressing edges of precipitation fronts. J. Chem. Phys. 120 (2004) 2413–2416.

De Lacy Costello B. P. J., Adamatzky A., Ratcliffe N. M., Zanin A., Purwins H. G., and Liehr A. The formation of Voronoi diagrams in chemical and physical systems: experimental findings and theoretical models. Int. J. Bifurcat. Chaos 14(7) (2004) 2187–2210.

De Lacy Costello B. and Adamatzky A. Experimental implementation of collision-based gates in Belousov–Zhabotinsky medium. Chaos Solitons Fractals 25 (2005) 535-544.

De Lacy Costello B. P. J., Jahan I., Adamatzky A., and Ratcliffe N. M. Chemical tesselations. Int. J. Bifurcat. Chaos 19 (2009) 619–622.

Delaunay B. Sur la sphère vide. Izv. Akad. Nauk SSSR Otd. Mat. Estestv. Nauk 7 (1934) 793–800.

Dexter S., Doyle P., and Gurevich Yu. Gurevich abstract state machines and Schönhage storage modification machines. J. Univers. Comput. Sci. 3 (1997) 279–303.

Dijkstra E. A. A note on two problems in connection with graphs. Numer. Math. 1 (1959) 269–271.

Diguet A., Guillermic R. M., Magome N., Saint-Jalmes A., Yong C., Yoshikawa C., and Baigl D. Photomanipulation of a droplet by the chromocapillary effect. Angew. Chem. Int. Ed. 48 (2009) 9281–9284.

Discover life. Physarum polycephalum. http://www.discoverlife.org/mp/20q?search=Physarum+polycephalum

Durham A. C. and Ridgway E. B. Control of chemotaxis in *Physarum polycephalum*. J. Cell Biol. 69 (1976) 218–223.

Epstein I. R. and Vanag V. K. Complex patterns in reactive microemulsions: self-organized nanostructures? Chaos 15 (2005) 047510.

Field R. J. and Noyes R. M. Oscillations in chemical systems. IV. Limit cycle behavior in a model of a real chemical reaction. J. Chem. Phys. 60 (1974) 1877–1884.

Fortune S. A sweepline algorithm for Voronoi diagrams. In Proc. 2nd Annu. Symp. Computational Geometry, Yorktown Heights, New York, 1986, pp. 313-322.

Fuerstman M. J., Deschatelets P., Kane R., Schwartz A., Kenis P. J. A., Deutch J. M., and Whitesides G. M. Solving mazes using microfluidic networks. Langmuir 19 (2003) 4714–4722.

Gabriel K. R. and Sokal R. R. A new statistical approach to geographic variation analysis. Syst. Zool. 18 (1969) 259–278.

Gacs P. and Levin L. A. Causal nets or what is a deterministic computation. STAN-CS-80-768, 1980.

Gawlitta W., Wolf K. V., Hoffmann H. U., and Stockem W. Studies on microplasmodia of Physarum polycephalum. I. Classification and locomotion behavior. Cell Tissue Res. 209 (1980) 71–86.

Górecki J., Yoshikawa K., and Igarashi Y. On chemical reactors that can count. J. Phys. Chem. A 107 (2003) 1664–1669.

Górecki J. and Górecka J. N. Multi-argument logical operations performed with excitable chemical medium. J. Chem. Phys. 124 (2006) 084101.

Górecki J. and Górecka J. N. Information processing with chemical excitations — from instant machines to an artificial chemical brain. Int. J. Unconv. Comput. 2 (2006) 321–336.

Górecki J., Górecka J. N., and Igarashi Y. Information processing with structured excitable medium. Nat. Comput. 8 (2009) 473–492.

Goss S., Aron. S., Deneubourg J. L., and Pasteels J. M. Self-organized shortcuts in the Argentine ant. Naturwissenschaften 76 (1989) 579–581.

Grigoriev D. Kolmogorov algorithms are stronger than Turing machines. Not. Sci. Semin. LOMI 60 (1976) 29–37 (in Russian). Engl. transl. J. Sov. Math. 14(5) (1980) 1445–1450.

Gurevich Y. On Kolmogorov machines and related issues. Bull. EATCS 35 (1988) 71–82.

Guttes E., Guttes S., and Rusch H. P. Morphological observations on growth and differentiation of Physarum polycephalum grown in pure culture. Dev. Biol. 3 (1961) 588–614.

Hagelbäck J. and Johansson S. J. A multi-agent potential field based bot for a full RTS game scenario. In Proc. AI and Interactive Digital Entertainment (AIIDE), 2009.

Hildebrandt A. A morphogen for the sporulation of Physarum polycephalum detected by cell fusion experiments. Exp. Cell Res. 167 (1986) 453–457.

Hulsmann N. and Wohlfarth-Bottermann K. E. Spatio-temporal relationships between protoplasmic streaming and contraction activities in plasmodial veins of Physarum polycephalum. Cytobiologie 17 (1978) 317–334.

Ing B. The Myxomycetes of Britain and Ireland: An Identification Handbook (Richmond Publishing Company, Slough, 1999).

Jaromczyk J. W. and Kowaluk M. A note on relative neighbourhood graphs. In Proc. 3rd Annu. Symp. Computational Geometry, 1987, pp. 233–241.

Jaromczyk J. W. and Toussaint G. T. Relative neighborhood graphs and their relatives. Proc. IEEE 80 (1992) 1502–1517.

Jarrett T. C., Ashton D. J., Fricker M., and Johnson N. F. Interplay between function and structure in complex networks. Phys. Rev. E 74 (2006) 026116.

Jones J. Characteristics of pattern formation and evolution in approximations of Physarum transport networks. Artif. Life 16 (2010) 127–153.

Jones J. An emergent pattern formation approach to dynamic spatial problems via quantitative front propagation and particle chemotaxis. Int. J. Unconv. Comput. 4 (2008) 341–374.

Jones J. Approximating the behaviours of Physarum polycephalum for the construction and minimisation of synthetic transport networks. In Unconventional Computation 2009 (Lect. Notes Comput. Sci. 5715) (Springer, New York, 2009), pp. 191–208.

Jones J. Passive vs active approaches in particle approximations of reaction–diffusion computing. Int. J. Nanotechnol. Mol. Comput. 1 (2009) 37–63.

Kakiuchi Y., Takahashi T., Murakami A., and Ueda T. Light irradiation induces fragmentation of the plasmodium, a novel photomorphogenesis in the true slime mold Physarum polycephalum: action spectra and evidence for involvement of the Phytochrome. Photochem. Photobiol. 73 (2001) 324-329.

Kaminaga A., Vanag V. K., and Epstein I. R. A reaction–diffusion memory device. Angew. Chem. Int. Ed. 45 (2006) 3087-3089.

Kamiya N. The protoplasmic flow in the myxomycete plasmodium as revealed by a volumetric analysis. Protoplasma 39 (1950) 3.

Kamiya N. Protoplasmic streaming. Protoplasmatologia 8 (1959) 1-199. Cited in [Tero et al. (2007)].

Katz E. and Privman V. Enzyme-based logic systems for information processing. Chem. Soc. Rev. 39 (2010) 1835–1857.

Kennedy B., Melhuish C., and Adamatzky A. Biologically inspired robots. In Electroactive Polymer (EAP) Actuators — Reality, Potential and Challenges, ed. by Bar-Cohen Y. (SPIE Press, Bellingham, Washington, 2001).

Kirkpatrick D. G. and Radke J. D. A framework for computational morphology. In Computational Geometry, ed. by Toussaint G. T. (North-Holland, Amsterdam, 1985), pp. 217-248.

Kitahata H., Aihara R., Magome N., and Yoshikawa K. Convective and periodic motion driven by a chemical wave. J. Chem. Phys. 116 (2002) 5666–5672.

Kitahata H. Spontaneous motion of a droplet coupled with a chemical reaction. Prog. Theor. Phys. Suppl. 161 (2006) 220–223.

Knuth D. E. The Art of Computer Programming, vol. 1: Fundamental Algorithms (Addison-Wesley, Reading, MA, 1968).

Kolmogorov A. N. On the concept of algorithm. Usp. Mat. Nauk 8(4) (1953) 175–176.

Kolmogorov A. N. and Uspensky V. A. On the definition of an algorithm. Usp. Mat. Nauk 13 (1958) 3–28 (in Russian). Engl. transl. ASM Transl. 21(2) (1963) 217–245.

Korohoda W., Shraideh Z., Baranowski Z., and Wohlfarth-Bottermann K. E. The blue-light reaction in plasmodia of *Physarum polycephalum* is coupled to respiration. Planta 158 (1983) 54–62.

Kruskal J. B. On the shortest subtree of a graph and the traveling problem. Proc. Am. Math. Soc. 7 (1956) 48–50.

Kuhnert L. A new photochemical memory device in a light sensitive active medium. Nature 319 (1986) 393.

Kuhnert L., Agladze K. L., and Krinsky V. I. Image processing using light-sensitive chemical waves. Nature 337 (1989) 244–247.

Lagzi I., Soh S., Wesson P. J., Browne K. P., and Grzybowski B. J. Maze solving by chemotactic droplets. J. Am. Chem. Soc. 132 (2010) 1198-119.

Lederman H., Macdonald J., Stefanovic D., and Stojanovic M. N. Deoxyribozyme-based three-input logic gates and construction of a molecular full adder. Biochemistry 45 (2006) 1194–1199.

Lyons R. and Peres Y. Probability on Trees and Networks, 1997. http://mypage.iu.edu/~rdlyons/prbtree/prbtree.html

Macdonald J., Li Y., Sutovic M., Lederman H., Pendri H., Lu W., Andrews B. L., Stefanovic D., and Stojanovic M. N. Medium scale integration of molecular logic gates in an automaton. Nano Lett. 6 (2006) 2598–2603.

Matsumoto K., Ueda T., and Kobatake Y. Propagation of phase wave in relation to tactic responses by the plasmodium of *Physarum polycephalum*. J. Theor. Biol. 122 (1986) 339–345.

Matsumoto K., Ueda T., and Kobatake Y. Reversal of thermotaxis with oscillatory stimulation in the plasmodium of Physarum polycephalum. J. Theor. Biol. 130 (1980) 175–182.

Matthiessen K. and Müller S. C. Chemically driven convection in the Belousov–Zhabotinsky reaction. Lect. Notes Phys. 464 (1996) 371–384.

Matula D. W. and Sokal R. R. Properties of Gabriel graphs relevant to geographical variation research and the clustering of points in the same plane. Geogr. Anal. 12 (1984) 205–222.

McAlister W. H. The diving and surface-walking behaviour of Dolomedes triton sexpunctatus (Araneida: Pisauridae). Anim. Behav. 8 (1959) 109–111.

Mebatsion H. K., Verboven P., Verlinden B. E., Ho Q. T., Nguyen T. A., and Nicolai B. M. Microscale modelling of fruit tissue using Voronoi tessellations. Comput. Electron. Agric. 52 (2006) 36–48.

Melhuish C., Adamatzky A., and Kennedy B. Biologically inspired robot. In SPIE 8th Annu. Int. Symp. Smart Structures and Materials, Newport Beach, CA, 2001.

Mills J. The nature of the extended analog computer. In Teuscher C., Nemenman I. M., and Alexander F. J. (eds.), Physica D. Special Issue: Novel Computing Paradigms: Quo Vadis? 237 (2008) 1235–1256.

Miyake Y., Tabata S., Murakami H., Yano M., and Shimizu H. Environment-dependent self-organization of positional information field in chemotaxis of Physarum Plasmodium. J. Theor. Biol. 178 (1996) 341–353.

Morgan A. P. The myxomecetes of the Miami Valley, Ohio. J. Cincinnati Soc. Nat. Hist. October 1892–January 1893. http://www.gutenberg.org/

files/29534/29534-h/29534-h.htm

Motoike I. N. and Yoshikawa K. Information operations with multiple pulses on an excitable field. Chaos Solitons Fractals 17 (2003) 455–461.

Murase Y., Maeda S., Hashimoto S., and Yoshida R. Design of a mass transport surface utilizing peristaltic motion of a self-oscillating gel. Langmuir 25 (2009) 483–489.

Nagai K., Sumino Y., Kitahata H., and Yoshikawa K. Mode selection in the spontaneous motion of an alcohol droplet. Phys. Rev. E 71 (2005) 065301.

Nakagaki T., Yamada H., and Ueda T. Modulation of cellular rhythm and photoavoidance by oscillatory irradiation in the Physarum plasmodium. Biophys. Chem. 82 (1999) 23–28.

Nakagaki T., Yamada H., and Toth A. Maze-solving by an amoeboid organism. Nature 407 (2000) 470.

Nakagaki T., Yamada H., and Toth A. Path-finding by tube morphogenesis in an amoeboid organism. Biophys. Chem. 92 (2001) 47-52.

Nakagaki T., Iima M., Ueda T., Nishiura Y., Saigusa T., Tero A., Kobayashi R., and Showalter K. Minimum-risk path finding by an adaptive amoeba network. Phys. Rev. Lett. 99 (2007) 068104.

Nakagaki T., Makoto I., Ueda T., Nishiura T., Saigusa T., Tero A., Kobayashi R., and Showalter K. Minimum-risk path finding by an adaptive amoebal network. Phys. Rev. Lett. 99 (2007) 68104.

Nesetril J., Milkova E., and Nesetrilova H. Otakar Boruvka on minimum spanning tree problem. Discrete Math. 233 (2001) 3–36.

Newton S. A., Ford N. C. Jr, Langley K. H., and Sattelle D. B. Laser light-scattering analysis of protoplasmic streaming in the slime mold Physarum polycephalum. Biochim. Biophys. Acta 496 (1977) 212–224.

Okabe A., Boots B., Sugihara K., and Chiu S. N. Spatial Tesselations (Wiley, Chichester, New York, 2000).

Paterson M. S. and Yao F. F. On nearest-neighbor graphs. In Proc. 19th. Int. Colloq. Automata Languages and Programming, 1992 (Lect. Notes Comput. Sci. 623), pp. 416–426.

Pointer G. Focus on People and Migration 2005. Chap. 3: The UKs Major Urban Areas (UK Statistics Authority, 2005). www.statistics.gov.uk

Preparata F. P. and Shamos M. I. Computational Geometry: An Introduction (Springer, New York, 1985).

Prim R. C. Shortest connection networks and some generalizations. Bell Syst. Tech. J. 36 (1957) 1389–1401.

Privman V., Pedrosa V., Melnikov D., Pita M., Simonian A., and Katz E. Enzymatic AND-gate based on electrode-immobilized glucose-6-phosphate dehydrogenase: towards digital biosensors and biochemical logic systems with low noise. Biosens. Bioelectron. 25 (2009) 695–701.

Rambidi N. G. Neural network devices based on reaction–diffusion media: an approach to artificial retina. Supramol. Sci. 5 (1998) 765–767.

Rambidi N. G., Shamayaev K. R., and Peshkov G. Yu. Image processing using light-sensitive chemical waves. Phys. Lett. A 298 (2002) 375–382.

Rapp P. E. The atlas of cellular oscillators. J. Exp. Biol. 81 (1979) 281–306.

Reyes D. R., Ghanem M. G., and George M. Glow discharge in micro fluidic chips for visible analog computing. Lab on a Chip 1 (2002) 113–116.

Ridgway E. B. and Durham C. H. Oscillations of calcium ion concentrations in *Physarum polycephalum*. J. Cell Biol. 69 (1976) 223–226.

Romanovskii Y. M. and Teplov V. A. The physical bases of cell movement. The mechanisms of self-organisation of amoeboid motility. Usp. Phys. 38 (1995) 521–542.

Saigusa T., Tero A., Nakagaki T., and Kuramoto Y. Amoebae anticipate periodic events. Phys. Rev. Lett. 100 (2008) 018101.

Sauer H. W., Babcock K. L., and Rusch H. P. Sporulation in Physarum polycephalum: a model system for studies on differentiation. Exp. Cell Res. 57 (1969) 319–327.

Schönhage A. Real-time simulation of multi-dimensional Turing machines by storage modification machines. Project MAC Technical Memorandum 37, MIT (1973).

Schönhage A. Storage modification machines. SIAM J. Comput. 9 (1980) 490–508.

Schreckenbach T., Walckhoff B., and Verfuerth C. Blue-light receptor in a white mutant of Physarum polycephalum mediates inhibition of spherulation and regulation of glucose metabolism. Proc. Natl. Acad. Sci. USA 78 (1981) 1009-1013.

Schreckenbach T. Phototaxis and photomorphogenesis in *Physarum polycephalum* plasmodia. In Blue Light Effects in Biological Systems, ed. by Senger H. (Springer, New York, 1984), pp. 464–475.

Schumann A. and Adamatzky A. Physarum spatial logic. In Proc. 11th Int. Symp. Symbolic and Numeric Algorithms for Scientific Computing, Timisoara, Romania, 26–29 September 2009.

Sedina-Nadal I., Mihaliuk E., Wang J., Perez-Munuzuri V., and Showalter K. Wave propagation in subexcitable media with periodically modulated excitability. Phys. Rev. Lett. 86 (2001) 1646-1649.

Shirakawa T. Private communication, February 2007.

Shirakawa T. and Gunji Y.-P. Computation of Voronoi diagram and collision-free path using the Plasmodium of Physarum polycephalum. Int. J. Unconv. Comput. 6 (2009) 79–88.

Shirakawa T., Adamatzky A., Gunji Y.-P., and Miyake Y. On simultaneous construction of Voronoi diagram and Delaunay triangulation by Physarum polycephalum. Int. J. Bifurcat. Chaos 19 (2009) 3109–3117.

Sielewiesiuk J. and Górecki J. Logical functions of a cross junction of excitable chemical media. J. Phys. Chem. A 105 (2001) 8189.

Skachek S., Adamatzky A., and Melhuish C. Manipulating objects by discrete excitable media coupled with contact-less actuator array: open-loop case. Chaos Solitons Fractals 26 (2005) 1377–1389.

Skachek S., Adamatzky A., and Melhuish C. Manipulating planar shapes by light-sensitive excitable medium: computational studies of closed loop systems. Int. J. Bifurcat. Chaos 16 (2006) 2333–2349.

Shvachko K. V. Different modifications of pointer machines and their computational power. In Proc. Symp. Mathematical Foundations of Computer

Science (MFCS 1991), pp. 426–435 (Lect. Notes Comput. Sci. 520).

Starona K. and Wojciech W. Light-induced transient increase of the activity of topoisomerase I in plasmodia of Physarum polycephalum. Int. J. Biochem. 24 (1992) 1717–1720.

Starostzik C. and Marwan W. A photoreceptor with characteristics of phytochrome triggers sporulation in the true slime mould Physarum polycephalum. FEBS Lett. 370 (1995) 146–148.

Steinbock O., Kettunen P., and Showalter K. Chemical wave logic gates. J. Phys. Chem. 100 (1996) 18970-18975.

Stephenson S. and Stempen H. Myxomeceters. A Handbook of Slime Molds (Timber Press, Portland, Oregon, 1994).

Stewart P. A. and Stewart B. T. Protoplasmic streaming and the fine structure of slime mold plasmodia. Exp. Cell. Res. 18 (1959) 374–377.

Stojanovic M. N., Mitchell Y. E., and Stefanovic D. Deoxyribozyme-based logic gates. J. Am. Chem. Soc. 124 (2002) 3555–3556.

Stojanovic M. N., Semova S., Kolpashchikov D., Macdonald J., Morgan C., and Stefanovic D. Deoxyribozyme-based ligase logic gates and their initial circuits. J. Am. Chem. Soc. 127 (2005) 6914–6915.

Sumino Y., Kitahata H., Yoshikawa K., Nagayama M., Nomura S.I., Magome N., and Mori Y. Chemosensitive running droplet. Phys. Rev. E 72 (2005) 041603.

Supowit K. J. The relative neighbourhood graph, with application to minimum spanning tree. J. ACM 30 (1988) 428–448.

Suter R. B., Rosenberg O., Loeb S., Wildman H., and Long J. Jr. Locomotion on the water surface: propulsive mechanisms of the fisher spider, Dolomedes triton. J. Exp. Biol. 200 (1997) 2523–2538.

Suter R. B. Cheap transport for fishing spiders: the physics of sailing on the water surface. J. Arachnol. 27 (1999) 489–496.

Suter R. B. and Wildman H. Locomotion on the water surface: hydrodynamic constraints on rowing velocity require a gait change. J. Exp. Biol. 202 (1999) 2771–2785.

Takahashi K., Uchida G., Hu Z., and Tsuchiya Y. Entrainment of the self-sustained oscillation in a *Physarum polycephalum* strand as a one-dimensionally coupled oscillator system. J. Theor. Biol. 184 (1997) 105–110.

Takamatsu A., Fujii T., and Endo I. Control of interaction strength in a network of the true slime mold by a microfabricated structure. BioSystems 55 (2000) 33-38.

Takamatsu A. Spontaneous switching among multiple spatio-temporal patterns in three-oscillator systems constructed with oscillatory cells of true slime mold. Physica D 223 (2006) 180-188.

Takamatsu A. Mobiligence in an amoeboid cell, plasmodium of Physarum polycephalum. In 2nd Int. Symp. Mobilgence, Awaji, Japan, 2007, pp. 48–51.

Tarassenko L. and Blake A. Analogue computation of collision-free paths. In: Proc. 1991 IEEE Int. Conf. Robotics and Automation, vol. 1, pp. 540–545.

Tarjan R. E. Reference machines require non-linear time to maintain disjoint sets. STAN-CS-77-603, 1977.

Terayama K., Honma H., and Kawarabayashi T. Toxicity of heavy metals and insecticides on slime mold Physarum polycephalum. J. Toxicol. Sci. 3 (1978) 293–303.

Tero A., Kobayashi R., and Nakagaki T. A coupled-oscillator model with a conservation law for the rhythmic amoeboid movements of plasmodial slime molds. Physica D 205 (2005) 125-135.

Tero A., Kobayashi R., and Nakagaki T. A mathematical model for adaptive transport network in path finding by true slime mold. J. Theor. Biol. 244 (2007) 553-564.

Teuscher C. and Adamatzky A. (eds.) Unconventional Computing 2005: From Cellular Automata to Wetware (Luniver Press, Beckington, 2005).

Tirosh R., Oplatka A., and Chet I. Motility in a 'cell sap' of the slime mold *Physarum Polycephalum*. FEBS Lett. 34 (1973) 40–42.

Tolmachiev D. and Adamatzky A. Chemical processor for computation of Voronoi diagram. Adv. Mater. Opt. Electron. 6 (1996) 191–196.

Toth R., Stone C., Adamatzky A., de Lacy Costello B., and Bull L. Experimental validation of binary collisions between wave-fragments in the photosensitive Belousov–Zhabotinsky reaction. Chaos Solitons Fractals 41 (2009) 1605–1615.

Toussaint G. T. The relative neighbourhood graph of a finite planar set. Pattern Recognition 12 (1980) 261–268.

Toussaint G. T. Some unsolved problems on proximity graphs. In 1st Workshop Proximity Graphs, Las Cruces, NM, December 1989. http://cgm.cs.mcgill.ca/~godfried/publications/openprox.pdf

Tsuda S., Aono M., and Gunji Y.-P. Robust and emergent Physarum-computing. BioSystems 73 (2004) 45–55.

Tsuda S., Zauner K. P., and Gunji Y. P. Robot control: from silicon circuitry to cells. In Biologically Inspired Approaches to Advanced Information Technology, ed. by Ijspeert A. J., Masuzawa T., and Kusumoto S. (Springer, New York, 2006), pp. 20–32.

Tsuda S., Zauner K. P., and Gunji Y. P. Robot control with biological cells. BioSystems 87 (2007) 215–223.

Tyson J. J. and Fife P. C. Target patterns in a realistic model of the Belousov-Zhabotinskii reaction. J. Chem. Phys. 73 (1980) 2224–2237.

Ueda T. and Kobatake Y. Contraction rhythm in the plasmodium of Physarum polycephalum: dependence of the period on the amplitude, temperature and chemical environment. Eur. J. Cell Biol. 23 (1980) 37–42.

Ueda T., Mori Y., and Kobatake Y. Patterns in the distribution of intracellular ATP concentration in relation to coordination of amoeboid cell behavior in Physarum polycephalum. Exp. Cell Res. 169 (1987) 191–201.

Ueda T. Pattern dynamics in cellular perception. Phase Transitions 45 (1993) 93–104.

Urquhart R. B. Algorithms for computation of relative neighbourhood graph. Electron. Lett. 16 (1980) 556–557.

Uspensky V. A. Kolmogorov and mathematical logic. J. Symb. Logic 57 (1992) 385–412.

van Emde Boas P. Space measures for storage modification machines. Inf. Process. Lett. 30 (1989) 103–110.

Wohlfarth-Bottermann K. E. and Block I. The pathway of photosensory transduction in Physarum polycephalum. Cell Biol. Int. Rep. 5 (1981) 365–373.

Wu Y., Vasquez D. A., Edwards B. E., and Wilder J. W. Convective chemical-wave propagation in the Belousov–Zhabotinsky reaction. Phys. Rev. E 51 (1995) 1119–1127.

Yamada H., Nakagaki T., Baker R. E., and Maini P. K. Dispersion relation in oscillatory reaction–diffusion systems with self-consistent flow in true slime mold. J. Math. Biol. 54 (2007) 745–760.

Yokoi H. and Kakazu Y. Theory and applications of autonomous machines based on the vibrating potential method. In Proc. Int. Symp. Distributed Autonomous Robotic Systems (IEEE, New York, 1992), pp. 31–38.

Yokoi H., Nagai T., Ishida T., Fujii M., and Iida T. Amoeba-like robots in the perspective of control architecture and morphology/materials. In Morpho-Functional Machines: The New Species, ed. by Hara F. and Pfeifer R. (Springer, Tokyo, 2003), pp. 99–129.

Yoshida R., Sakai T., Hara Y., Maeda S., Hashimoto S., Suzuki D., and Murase Y. Self-oscillating gel as novel biomimetric materials. J. Control Release 140 (2009) 186–193.

Yoshikawa K., Motoike I. M., Ichino T., Yamaguchi T., Igarashi Y., Gorecki J., and Gorecka J. N. Basic information processing operations with pulses of excitation in a reaction–diffusion system. Int. J. Unconv. Comput. 5 (2009) 3–37.

Index

www.ingramcontent.com/pod-product-compliance
Lightning Source LLC
Chambersburg PA
CBHW050548190326
41458CB00007B/1962